TOPICS IN
STATISTICAL
MECHANICS

Imperial College Press Advanced Physics Texts

Forthcoming:

Elements of Quantum Information
  by Martin Plenio & Vlatko Vedral

Physics of the Environment
  by Andrew Brinkman

Modern Astronomical Instrumentation
  by Adrian Webster

Symmetry, Groups and Representations in Physics
  by Dimitri Vvedensky & Tim Evans

Imperial College Press Advanced Physics Texts – Vol. 3

TOPICS IN
STATISTICAL
MECHANICS

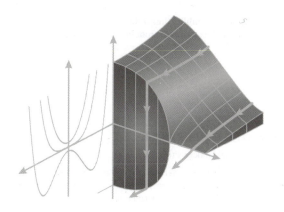

Brian Cowan

Royal Holloway, University of London, UK

Imperial College Press

Published by

Imperial College Press
57 Shelton Street
Covent Garden
London WC2H 9HE

Distributed by

World Scientific Publishing Co. Pte. Ltd.
5 Toh Tuck Link, Singapore 596224
USA office: 27 Warren Street, Suite 401-402, Hackensack, NJ 07601
UK office: 57 Shelton Street, Covent Garden, London WC2H 9HE

British Library Cataloguing-in-Publication Data
A catalogue record for this book is available from the British Library.

ISBN   1-86094-564-3
ISBN   1-86094-569-4 (pbk)

Typeset by Stallion Press
Email: enquiries@stallionpress.com

Printed in Singapore by World Scientific Printers (S) Pte Ltd

PREFACE

David Goodstein. . . . . . .

'Ludwig Boltzmann, who spent much of his life studying statistical mechanics, died in 1906, by his own hand. Paul Ehrenfest, carrying on the work, died similarly in 1933. Now it is our turn to study statistical mechanics.

Perhaps it will be wise to approach the subject cautiously.'

— in *States of Matter*, 1975, Dover N.Y.

Statistical Mechanics, more than any other branch of Physics, is beset with problems of methodology and presentation. Philosophers have argued over the meaning of probability, particularly when applied to a single "event". Mathematicians have neatly side-stepped the issue by stripping away physical interpretation and treating probability simply as a "measure" accompanied by a set of rules. But disengaging from reality in this way is of scant use to Physics. For Physicists, probability and the statistical method continue to cause their share of anguish. The statistical method was a contributory factor to Boltzmann's suicide and possibly to that of Paul Ehrenfest. Even today the conundrums of Quantum Mechanics have aspects of probability at their heart.

In Statistical Mechanics, the operational approach of the mathematicians is paralleled by the Information Theory approach of E. T. Jaynes. True, this method is championed by some outstanding pedagogues, but I confess to finding it distinctly unappealing. Certainly it might be an expedient way to obtain results, but (to my mind) it obscures *understanding*. And understanding is central to the Physicist's endeavours. By contrast, the ensemble formalism of scholars such as T. L. Hill provides maximum clarity and physical meaning. Some might find this overly formal; I find it distinctly appealing.

The second structural issue is the relationship between Statistical Mechanics and the older discipline of Thermodynamics. Should these be developed independently or are they better treated in a unified manner? Landau was a strong supporter of the latter view and I have mostly been persuaded by his arguments. At the undergraduate level this philosophy benefited from the magnificent exposition of Reif (also his volume in the Berkley Physics series). It is probable that this had a major impact on the intellectual formation of many of today's professional physicists. It was certainly an ideal preparation for study of the appropriate Landau and Lifshitz volume(s). Nevertheless there is no doubt that classical thermodynamics has its strengths. Most important is the model-independence of its results. Aspects of its abstract formalism may deter some and indeed parts of its deeper logic may be obscure: witness the gradual conversion to the Carathéodory viewpoint by Zemansky through the many editions of his Thermodynamics textbook. These days the logical aspects of Classical Thermodynamics have been elegantly clarified and reformulated by Callen, but this is probably not appropriate at an undergraduate level.

Regardless of the deeper philosophical issues, this second matter has been mostly resolved through necessity. The increasing pressure on the undergraduate curriculum means that in most UK universities there is no longer the space available to present self-contained courses on Classical Thermodynamics and Statistical Mechanics. Indeed this is the case at Royal Holloway University of London. However the introduction of the four-year undergraduate integrated Masters degree, the M.Sci or M.Phys, has helped and allowed the incorporation of some more advanced material into the degree programmes.

Within the University of London, King's, Queen Mary, Royal Holloway, and University Colleges collaborate in the joint teaching of the fourth year of the M.Sci degree, with lectures held in central London. This has allowed a wide range of courses to be provided, many of which are at the cutting edge of the subject. And at the same time this has given the opportunity for a more detailed coverage of some material latterly squeezed out of the traditional three-year B.Sc. degrees.

This book has arisen out of such an intercollegiate course in Statistical Mechanics that I have taught to M.Sci fourth year students over the last ten or so years. Such intercollegiate teaching presents its own challenges. At the completion of their third year, students are required to have reached a common standard and level for embarking on the intercollegiate fourth year, although presentation and flavour will have been different in each college. This is particularly the case in the area of thermal physics, for the reasons outlined in the first few paragraphs above. And, because of this, the learning material for this course provided to students included some more elementary "foundation" material, albeit presented from a slightly more mature standpoint. Approximately half the material of Chapters One and Two falls into this category.

This was a "paperless" course, whereby students were provided with lecture notes and other learning material on the web. That included the lecture calendar, problem exercises, etc., all of which students could access remotely from their home college or from elsewhere. I was alerted to the wider appeal of the course material when I started to receive requests from students, outside London and around the world, for answers to problem exercises (often with deadlines!) and queries about unclear portions of the notes. This became strikingly poignant on occasions when the college web service was interrupted. Then I would often find frantic email enquiries asking where the material had gone. So when I was approached by Imperial College Press to consider producing a book version of the course, I eventually agreed. This book comprises a re-working of the course material, incorporating changes and suggestions from the various cohorts of students, colleagues and reviewers commissioned by ICP.

Electromagnetic units have traditionally been the cause of many problems to students learning their Physics from a range of sources. It was expected that such difficulties would have been eliminated through the adoption of the SI system. But there has been resistance. Kittel's Solid State Physics book has settled on a compromise whereby many equations are quoted in both their Gaussian and SI forms. I have adopted, almost exclusively, SI units although I am unhappy about aspects of the B-H controversy that this can lead to. However readers should note that I use the symbol M to represent total magnetic moment rather than magnetic moment per unit volume. I also confess to the eccentricity of representing the complex dynamical susceptibility as $\chi(\omega) = \chi'(\omega) + i\chi''(\omega)$ rather than the more common complex conjugate form.

I am indebted to many people. Louise, my wife and Abigail, my daughter have become used to an academic husband and father who is so often absent — mentally, if not physically. My teachers Michael Richards and Bill Mullin stimulated what was to become an ongoing obsession with this branch of physics. Their endeavours prepared me for study of the Landau and Lifshitz volumes; I continue to be thrilled with the incisive clarity and remarkable depth of those books. In my early teaching career I learned much from Roland Dobbs and from Mike Hoare. More recently, Bob Jones has repeatedly demonstrated his encyclopaedic knowledge of statistical mechanics in responding to my obscure questions. His gentle approach has often helped in formulating a student-friendly treatment of a difficult topic. And as ever, John Saunders remains my "critical friend"; he continues to be a constant source of inspiration in so many ways. Above all, I am grateful to the many students whom I have had the privilege of teaching. Their observations, questions, and indeed objections, have helped me to clarify my own views while eliminating sloppy argumentation. Nevertheless, these acknowledgements in no way imply an abrogation of my pedagogical duties. Errors, both of omission and of commission, remain my responsibility alone.

I dedicate this book to the memory of my father, Stanley Cowan, who died in February 1997. He was an engineer of rare ingenuity and a man of exceptional patience. He bore his terminal illness with a serenity that humbled those who knew him. He strongly believed in the power of mathematics in the solution of problems and he imbued me with that belief. He encouraged me to ask questions and although a robust debater, he kept an open mind to the end.

Brian Cowan
December 2004.

CONTENTS

CHAPTER ONE

THE METHODOLOGY OF STATISTICAL MECHANICS

This chapter provides an overview of statistical mechanics and thermodynamics. Although the discussion is reasonably self-contained, it is assumed that this is not the reader's first exposure to these subjects. More than many other branches of physics, these topics are treated in a variety of different ways in the textbooks in common usage. The aim of the chapter is thus to establish the perspective of the book, the standpoint from which the topics in the following chapters will be viewed. The textbook by Bowley[1] is recommended to the novice as a clearly-argued introduction to the subject; indeed all readers will find many of the examples of this book treated by that book in a complementary fashion.

1.1. Terminology and Methodology

1.1.1. *Approaches to the subject*

Thermodynamics is the study of the relationship between *macroscopic* properties of systems such as temperature, volume, pressure, magnetisation, compressibility, etc. *Statistical Mechanics* is concerned with understanding how the various macroscopic properties arise as a consequence of the *microscopic* nature of the system. In essence it makes macroscopic deductions from microscopic models.

The power of thermodynamics, as formulated in the traditional manner (e.g. Zemansky[2]) is that its deductions are quite general; they do not rely, for their validity, on the microscopic nature of the system. Einstein

1

expressed this quite impressively when he wrote[3]

> "*A theory is the more impressive the greater the simplicity of its premises is, the more different kinds of things it relates, and the more extended is its area of applicability. Therefore the deep impression which classical thermodynamics made upon me; it is the only physical theory of universal content concerning which I am convinced that, within the framework of the applicability of its basic concepts, will never be overthrown.*"

On the other hand statistical mechanics, as conventionally presented (e.g. Hill[4]) is system-specific. One starts from particular microscopic models, say the Debye model for a solid, and derives macroscopic properties such as the thermal capacity. It is true that statistical mechanics will give relationships between the various macroscopic properties of a system, but they will only apply to the system/model under consideration. Results obtained from thermodynamics, on the other hand, are model-independent and general.

Traditionally thermodynamics and statistical mechanics were developed as independent subjects, with "bridge equations" making the links between the two. Alternatively the subjects can be developed together, where the Laws of Thermodynamics are justified by microscopic arguments. This reductionist view was adopted by Landau. In justification he wrote[5]:

> "*Statistical physics and thermodynamics together form a unit. All the concepts and quantities of thermodynamics follow most naturally, simply and rigorously from the concepts of statistical physics. Although the general statements of thermodynamics can be formulated non-statistically, their application to specific cases always requires the use of statistical physics.*"

The contrast between the views of Einstein and Landau is apparent. The paradox, however, is that so much of Einstein's work was reductionist and microscopic in nature whereas Landau was a master of the macroscopic description of phenomena. This book considers thermodynamics and statistical mechanics in a synthetic manner; in that respect it follows more closely the Landau approach.

1.1.2. *Description of states*

The state of a system described at the macroscopic level is called a *macrostate*. Macrostates are described by a relatively few variables such as temperature, pressure and volume.

The state of a system described at the microscopic level is called a *microstate*. Microstates are described by a very large number of variables. Classically you would need to specify the position and momentum of each particle in the system. Using a quantum-mechanical description you would have to specify all the quantum numbers of the entire system. So whereas a macrostate might be described by under ten variables, a microstate will be described by over 10^{23} variables.

The fundamental methodology of statistical mechanics involves applying probabilistic arguments about microstates, regarding macrostates as statistical averages. It is because of the very large number of particles involved that the mean behaviour corresponds so well to the observed behaviour — that fluctuations are negligible. Recall that for N particles the likely fractional deviation from the mean will be $1/\sqrt{N}$. So if you calculate a mean pressure of mole of air at one atmosphere, it is likely to be correct to within $\sim 10^{-12}$ of an atmosphere. It is also because of the large number of particles involved that the mean, observed behaviour corresponds to the most probable behaviour — the *mean* and the *mode* of the distribution differ negligibly.

Statistical Mechanics is slightly unusual in that its formalism is easier to understand using quantum mechanics for descriptions at the microscopic level rather than classical mechanics. This is because the idea of a quantum state is familiar, and often the quantum states are discrete. It is more difficult to enumerate the states of a classical system; it is best done by analogy with the quantum case. The classical case will be taken up in Sec. 1.6.

1.1.3. *Extensivity and the thermodynamic limit*

The thermodynamic variables describing macrostates fall into two classes. Quantities such as energy E, number of particles N and volume V, which add when two similar systems are combined, are known as *extensive* variables. And quantities such as pressure p and temperature T, which remain independent when similar systems are combined, are known as *intensive* variables.

As we have argued, thermodynamics concerns itself with the behaviour of "large" systems and usually, for these, finite-size effects are not of interest. Thus, for example, unless one is concerned specifically with surface phenomena, it will be sensible to focus on systems that are sufficiently large that the surface contribution to the energy will be negligible when compared to the volume contribution. This is possible because the energy of interaction between atoms is usually sufficiently short-ranged. In general one will consider properties in the limit $N \to \infty$, $V \to \infty$ while N/V remains constant. This is called the *thermodynamic limit*.

It should be apparent that true extensivity of a quantity such as energy arises only in the thermodynamic limit.

[Gravitation is a long-range force. It is apparent that gravitational energy is not truly extensive; this is examined in Problem 1.2. This non-extensivity makes for serious difficulties when trying to treat the statistical thermodynamics of gravitating systems. A generalisation of the definition of entropy to accommodate non-extensive systems was proposed by Tsallis in 1989. An account this is contained in the review article by Tsallis[6] and subsequent articles in the same journal.]

1.2. The Fundamental Principles

1.2.1. *The laws of thermodynamics*

Thermodynamics, as a logical structure, is built on its four assumptions or laws (including the zeroth).

Zeroth Law: **If system A is in equilibrium with system B and with system C then system B is in equilibrium with system C.** Equilibrium here is understood in the sense that when two systems are brought into contact then there is no change. This law was formalised after the first three laws, but because it was believed to be more fundamental it was called the *Zeroth* Law. The Zeroth Law recognises the existence of states of equilibrium and it points us to the concept of temperature, a non-mechanical quantity that can label (and order) equilibrium states.

First Law: **The internal energy of a body can change by the flow of heat or by doing work**

$$\Delta E = \Delta Q + \Delta W. \tag{1.1}$$

Here ΔQ is the energy increase as a consequence of heat flow and ΔW is the energy increase resulting from work done. We usually understand this as a statement about the conservation of energy. But in its historical context the law asserted that as well as the familiar mechanical form of energy, heat also was a form of energy. Today we understand this as the kinetic energy of the constituent particles of a system; in earlier times the nature of heat was unclear.

Note that some older books adopt the *opposite sign* for ΔW; they consider ΔW to be the work done *by* the system rather than the work done *on* the system.

Second Law (this law has many formulations): **Heat flows from hot to cold, or It is not possible to convert** *all* **heat energy to work**. These statements have the great merit of being reflections of common experience. There are other formulations such as the Carathéodory statement (see the books by Adkins[7] or Pippard[8]): **In the neighbourhood of any equilibrium state of a thermally isolated system there are states which are inaccessible**, and the entropy statement (see Callen's book[9]): **There is an extensive quantity, which we call entropy, which never decreases in a physical process**. The claimed virtue of the Carathéodory statement is that it leads more rapidly to the important thermodynamic concepts of temperature and entropy: this at the expense of common experience. But if this is believed to be a virtue then one may as well go the "whole hog" and adopt the Callen statement. A major exercise in classical thermodynamics is proving the equivalence of the various statements of the Second Law. In whatever form, the Second Law leads to the concept of entropy and the *quantification* of temperature (the Zeroth Law just gives an *ordering*: A is hotter than B). And it tells us there is an absolute zero of temperature.

Third Law: **The entropy of a body tends to zero as the temperature tends to absolute zero**. The Third Law will be discussed in Sec. 1.7. We shall see that it arises as a consequence of the quantum behaviour of matter at the microscopic level. However we see immediately that the Third Law is telling us there is an absolute zero of entropy.

An even more fundamental aspect of the Zeroth Law is the fact of the *existence* of equilibrium states. If systems did not exist in states of equilibrium then there would be no macrostates and no hope of description in terms of small numbers of variables. Then there would be no discipline of thermodynamics and phenomena would have to be discussed solely in terms of their intractable microscopic descriptions. Fortunately this is not the case; the existence of states of equilibrium allows our simple minds to make some sense of a complex world.

1.2.2. *Probabilistic interpretation of the First Law*

The First Law discusses the way the energy of a system can change. From the statistical standpoint we understand the energy of a macroscopic system as the *mean* value since the system can exist in a large number of different microstates. If the energy of the jth microstate is E_j and the probability of occurrence of this microstate is P_j then the (mean) energy of the system is given by

$$E = \sum_j P_j E_j. \tag{1.2}$$

The differential of this expression is

$$dE = \sum_j P_j dE_j + \sum_j E_j dP_j; \tag{1.3}$$

this indicates the energy of the system can change if the energy levels E_j change or if the probabilities P_j change.

The first term $\sum_j P_j dE_j$ relates to the change in the energy levels dE_j. This term is the mean energy change of the microstates and we shall show below that this corresponds to the familiar "mechanical" energy, the work done on the system. The second term $\sum_j E_j dP_j$ is a consequence of the change in probabilities or occupation of the energy states. This is fundamentally probabilistic and we shall see that it corresponds to the heat flow into the system.

In order to understand that the first term corresponds to the work done, let us consider a pV system. We shall see (Sec. 2.1.3) that the energy levels depend on the size (volume) of the system:

$$E_j = E_j(V)$$

so that the change in the energy levels when the volume changes is

$$dE_j = \frac{\partial E_j}{\partial V} dV.$$

Then

$$\sum_j P_j dE_j = \sum_j P_j \frac{\partial E_j}{\partial V} dV.$$

We are assuming that the change in volume occurs at constant P_j, then

$$\sum_j P_j dE_j = \sum_j \frac{\partial}{\partial V} P_j E_j dV$$

$$= \frac{\partial}{\partial V} E dV.$$

But we identify $\partial E/\partial V = -p$, so that

$$\sum_j P_j dE_j = -p\,dV. \tag{1.4}$$

And thus we see that the term $\sum_j P_j dE_j$ corresponds to the work done on the system. Then the term $\sum_j E_j dP_j$ corresponds to the energy increase of the system that occurs when no work is done; this is what we understand as heat flow.

We have seen, in this section, that the idea of heat arises quite logically from the probabilistic point of view.

1.2.3. *Microscopic basis for entropy*

By contrast to macroscopic thermodynamics, statistical mechanics is built on a single assumption, which we will call the Fundamental Postulate of statistical mechanics. We shall see how the Laws of thermodynamics may be understood in terms of this Fundamental Postulate, which dates back to Boltzmann. The Fundamental Postulate states: **All microstates of an isolated system are equally likely**. Note, in particular, that an isolated system will have fixed energy E, volume V and number of particles N (extensive quantities). We conventionally denote by $\Omega(E, V, N)$ the number of microstates in a given macrostate (E, V, N). Then from the Fundamental Postulate it follows that the probability of a given macrostate is proportional to the number of microstates corresponding to it $\Omega(E, V, N)$:

$$P \propto \Omega(E, V, N). \tag{1.5}$$

If we understand the observed equilibrium state of a system as the most probable macrostate, then it follows from the Fundamental Postulate that the equilibrium state corresponds to the macrostate with the largest number of microstates. We are saying that Ω is maximum for an equilibrium state.

Since Ω for two isolated systems is multiplicative, it follows that the *logarithm* of Ω is additive. In other words $\ln \Omega$ is an *extensive* quantity.

We define *entropy S* as

$$S = k \ln \Omega. \tag{1.6}$$

At this stage k is simply a constant; later we will identify it as Boltzmann's constant.

Since the logarithm is a monotonic function it follows that the equilibrium state will have maximal entropy. So we immediately obtain the

Second Law. And now we understand the Second Law from the microscopic point of view; it is hardly more than the tautology "we are most likely to observe the most probable state"!

1.3.   Interactions — The Conditions for Equilibrium

When systems interact their states will often change, the composite system evolving to a state of equilibrium. We shall investigate what determines the final state. We will see that quantities such as temperature emerge in the description of these equilibrium states.

1.3.1.   *Thermal interaction — Temperature*

Let us allow two systems to exchange energy without changing volume or numbers of particles. In other words we allow thermal interaction only; the systems are separated by a *diathermal* wall.

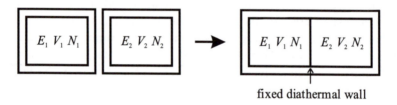

fixed diathermal wall

Fig. 1.1.   Thermal interaction.

Now

$$\Omega_1 = \Omega_1(E_1, V_1, N_1)$$
$$\Omega_2 = \Omega_2(E_2, V_2, N_2)$$

and V_1, N_1, V_2, N_2 are all fixed.

The energies E_1 and E_2 can vary subject to the restriction $E_1 + E_2 = E_0$, a constant.

Our problem is this: after the two systems are brought together what will be the equilibrium state? We know that the systems will exchange energy, and they will do this so as to maximise the total number of microstates for the composite system.

For different systems the Ωs multiply so that we have

$$\Omega = \Omega_1(E_1)\Omega_2(E_2) \tag{1.7}$$

— we can ignore V_1, N_1, V_2, N_2 as they do not change.

The systems will exchange energy so as to maximise Ω. Writing

$$\Omega = \Omega_1(E)\Omega_2(E_0 - E)$$

we allow the systems to vary E so that Ω is a maximum:

$$\frac{\partial \Omega}{\partial E} = \frac{\partial \Omega_1}{\partial E}\Omega_2 - \Omega_1\frac{\partial \Omega_2}{\partial E} = 0$$

or

$$\frac{1}{\Omega_1}\frac{\partial \Omega_1}{\partial E} = \frac{1}{\Omega_2}\frac{\partial \Omega_2}{\partial E}$$

or

$$\frac{\partial \ln \Omega_1}{\partial E} = \frac{\partial \ln \Omega_2}{\partial E}.$$

(Note the natural occurrence of the *logarithm* of Ω.)

But from the definition of entropy, $S = k \ln \Omega$, we see this means that the equilibrium state is characterised by

$$\frac{\partial S_1}{\partial E} = \frac{\partial S_2}{\partial E}. \tag{1.8}$$

In other words, when the systems have reached equilibrium the quantity $\partial S/\partial E$ of system 1 is equal to $\partial S/\partial E$ of system 2. This is the condition for equilibrium when systems exchange only thermal energy.

Clearly $\partial S/\partial E$ must be related to the temperature of the system. Since

$$\Delta S = \left(\frac{\partial S_1}{\partial E} - \frac{\partial S_2}{\partial E}\right)\Delta E_1 \geq 0 \tag{1.9}$$

this means that

$$E_1 \text{ increases if } \frac{\partial S_1}{\partial E} > \frac{\partial S_2}{\partial E}$$

$$E_1 \text{ decreases if } \frac{\partial S_1}{\partial E} < \frac{\partial S_2}{\partial E}$$

so energy flows from systems with small $\partial S/\partial E$ to systems with large $\partial S/\partial E$.

Since we know that heat flows from hot systems to cold systems, we therefore identify

$$\text{High } T \equiv \text{Low } \frac{\partial S}{\partial E}$$

$$\text{Low } T \equiv \text{High } \frac{\partial S}{\partial E}.$$

There is thus an *inverse* relation between $\partial S/\partial E$ and temperature.

We define *statistical temperature* by

$$\frac{1}{T} = \frac{\partial S}{\partial E}.$$ (1.10)

When applied to the ideal gas this will give us the result

$$pV = NkT.$$ (1.11)

And it is from this we conclude that the statistical temperature corresponds to the intuitive concept of temperature as measured by an ideal gas thermometer. Furthermore the *scale* of temperatures will agree with the Kelvin scale (ice point at 273.18 K) when the constant k in the definition of S is identified with Boltzmann's constant.

When the partial derivative $\partial S/\partial E$ is evaluated, N and V are constant. So the only energy flow is heat flow. Thus the equation defining statistical temperature can also be written as

$$\Delta Q = T\Delta S.$$ (1.12)

We can now write the energy conservation expression for the First Law:

$$\Delta E = \Delta Q + \Delta W$$ (1.13)

as

$$\Delta E = T\Delta S - p\Delta V \quad \text{(for } pV \text{ systems)}.$$ (1.14)

1.3.2. *Volume change — Pressure*

We now allow the volumes of the interacting systems to vary as well, subject to the total volume being fixed. Thus we consider two systems separated by a *movable* diathermal wall.

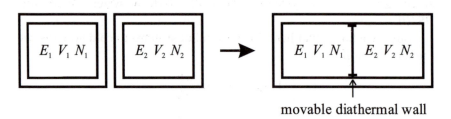

movable diathermal wall

Fig. 1.2.   Mechanical and thermal interaction.

The constraints on this system are

$$E_1 + E_2 = E_0 = \text{const.}$$
$$V_1 + V_2 = V_0 = \text{const.},$$

while N_1 and N_2 are individually fixed.

Maximising the entropy with respect to both energy flow and volume change then gives the two conditions

$$\frac{\partial S_1}{\partial E} = \frac{\partial S_2}{\partial E}$$

$$\frac{\partial S_1}{\partial V} = \frac{\partial S_2}{\partial V}. \tag{1.15}$$

The first of these, we know, gives the equality of temperature at equilibrium:

$$T_1 = T_2.$$

What does the second relation tell us? What is $\partial S / \partial V$? This may be found by rearranging the differential expression for the First Law:

$$dE = T dS - p dV.$$

This may be rewritten as

$$dS = \frac{1}{T} dE + \frac{p}{T} dV \tag{1.16}$$

so just as we identified

$$\left. \frac{\partial S}{\partial E} \right|_V = \frac{1}{T},$$

so we now identify

$$\left. \frac{\partial S}{\partial V} \right|_E = \frac{p}{T}. \tag{1.17}$$

Thus the condition that $\partial S / \partial V$ be the same for both systems means that p/T must be the same. But we have already established that T is the same so the new information is that at equilibrium the pressures are equalised:

$$p_1 = p_2. \tag{1.18}$$

[A paradox arises if the movable wall is not diathermal: that is, if it is thermally isolating. Then one would conclude, from an analysis similar to that above, that while p/T becomes equalised for the two sides, T does not. On the other hand, a purely mechanical argument would say that the pressures p should become equal. The paradox is resolved when one appreciates that without a flow of heat, thermodynamic equilibrium is not possible and so the entropy maximum principle is not applicable. Thus p/T will not be equalised. This issue is discussed in greater detail by Callen.[9]]

1.3.3. *Particle interchange — Chemical potential*

Let us keep the volumes of the two systems fixed, but allow particles to traverse the immobile diathermal wall.

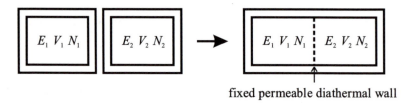

fixed permeable diathermal wall

Fig. 1.3.   Heat and particle exchange.

The constraints on this system are

$$E_1 + E_2 = E_0 = \text{const.}$$
$$N_1 + N_2 = N_0 = \text{const.},$$

while V_1 and V_2 are individually fixed.

Maximising the entropy with respect to both energy flow and particle flow then gives the two conditions

$$\frac{\partial S_1}{\partial E} = \frac{\partial S_2}{\partial E}$$

$$\frac{\partial S_1}{\partial N} = \frac{\partial S_2}{\partial N}.$$

The first of these, we know, gives the equality of temperature at equilibrium:

$$T_1 = T_2.$$

What does the second relation tell us? What is $\partial S/\partial N$? This may be found from the First Law in its extended form:

$$dE = TdS - pdV + \mu dN \qquad (1.19)$$

where μ is the chemical potential. This may be rewritten as

$$dS = \frac{1}{T}dE + \frac{p}{T}dV - \frac{\mu}{T}dN$$

so that we may identify

$$\left.\frac{\partial S}{\partial N}\right|_{E,V} = -\frac{\mu}{T}. \tag{1.20}$$

Thus the condition that $\partial S/\partial N$ be the same for both systems means that μ/T must be the same. But we have already established that T is the same so the new information is that at equilibrium the chemical potentials are equalised:

$$\mu_1 = \mu_2.$$

We see that just as pressure drives volume changes, chemical potential drives particle flow. And arguments similar to those of Sec. 1.3.1 indicate that particles flow from high values of μ to low values of μ.

1.3.4. *Thermal interaction with the rest of the world — The Boltzmann factor*

For an isolated system all microstates are equally likely; this is our Fundamental Postulate. It follows that the probability of the occurrence of a given *microstate* is given by

$$P_j = \frac{1}{\Omega}. \tag{1.21}$$

But what about a non-isolated system? What can we say about the occurrence of microstates of such a system? Here the probability of a microstate will depend on properties of the surroundings.

In effect we are seeking an extension of our Fundamental Postulate. We shall see how we can use the Postulate itself to effect its extension.

We consider a system interacting with its surroundings through a fixed diathermal wall; this non-isolated system can exchange thermal energy with its surroundings. We ask the question "what is the probability of this non-isolated system being in a given microstate?" This system will have its temperature determined by its environment.

We shall idealise the "rest of the world" by a heat bath at constant temperature. We shall regard the bath plus our system of interest as isolated — so to *this* we can apply the Fundamental Postulate. In this way, we shall be able to find the probability that the system of interest is in a particular microstate. This is the "wine bottle in the swimming pool" model of Reif.[10]

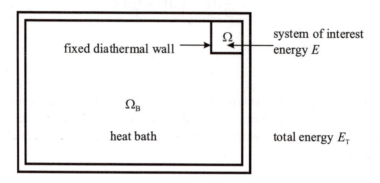

Fig. 1.4.   Thermal interaction with the rest of the world.

The Ωs multiply, thus

$$\Omega_T = \Omega_B \times \Omega$$

$$\uparrow \qquad \uparrow \qquad \uparrow$$

Total   Bath   System of interest

or

$$\Omega_T = \Omega_B(E_T - E)\Omega(E).$$

Now the Fundamental Postulate tells us that the probability the system of interest has energy E is proportional to the number of microstates of the composite system that correspond to that energy partition

$$P(E) \propto \Omega_B(E_T - E)\Omega(E).$$

But here $\Omega(E) = 1$ since we are looking at a given *microstate* of energy E — there is one microstate. So

$$P(E) \propto \Omega_B(E_T - E).$$

It depends solely on the bath. In terms of entropy, since $S = k \ln \Omega$,

$$P(E) \propto e^{S(E_T - E)/k} \tag{1.22}$$

where S is the entropy of the *bath*. This type of expression, where probability is expressed in terms of entropy is an inversion of the usual usage where entropy and other thermodynamic properties are found in terms of probabilities. This form was much used by Einstein in his treatment of fluctuations.

Now the subsystem is very small compared with the bath; $E \ll E_T$. So we can perform a Taylor expansion of S:

$$S(E_T - E) = S(E_T) - E\frac{\partial S}{\partial E} + \cdots$$

but

$$\frac{\partial S}{\partial E} = \frac{1}{T},$$

the temperature of the bath, so that

$$S(E_T - E) = S(E_T) - \frac{E}{T}$$

assuming we can ignore the higher terms. Then

$$P(E) \propto e^{S(E_T)/k}e^{-E/kT}.$$

But the first term $e^{S(E_T)/k}$ is simply a constant, so we finally obtain the probability

$$P(E) \propto e^{-E/kT}. \tag{1.23}$$

This is the probability that a system in equilibrium at a temperature T will be found in a microstate of energy E. The exponential factor $e^{-E/kT}$ is known as the Boltzmann factor, the Boltzmann distribution function or the canonical distribution function.

The Boltzmann factor is a key result. Feynman says[11]

"This fundamental law is the summit of statistical mechanics, and the entire subject is either a slide-down from the summit, as the principle is applied to various cases, or the climb-up to where the fundamental law is derived and the concepts of thermal equilibrium and temperature clarified".

1.3.5. *Particle and energy exchange with the rest of the world — The Gibbs factor*

We now consider an extension of the Boltzmann factor to account for microstates where the number of particles may vary. Our system here can exchange both energy and particles with the rest of the world. The microstate of our system of interest is now specified by a given energy and a given number of particles. Our question will be "what is the probability that the system of interest will be found with energy E and N particles?"

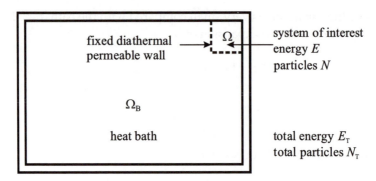

Fig. 1.5. Particle and energy exchange with the rest of the world.

In this case Ω_T is a function of both E and N.

$$\Omega_T = \Omega_B(E_T - E, N_T - N)\Omega(E, N).$$

Now the Fundamental Postulate tells us that the probability the system of interest has energy E and N particles is proportional to the number of microstates of the composite system that correspond to that energy and particle partition

$$P(E, N) \propto \Omega_B(E_T - E, N_T - N)\Omega(E, N).$$

But as before, $\Omega(E, N) = 1$ since we are looking at a single microstate. So

$$P(E, N) \propto \Omega_B(E_T - E, N_T - N).$$

It depends solely on the bath. In terms of entropy, since $S = k \ln \Omega$,

$$P(E, N) \propto e^{S(E_T - E, N_T - N)/k} \qquad (1.24)$$

where S is the entropy of the *bath*.

Now the subsystem is very small compared with the bath; $E \ll E_T$ and $N \ll N_T$. So, as before, we can do a Taylor expansion of S:

$$S(E_T - E, N_T - N) = S(E_T, N_T) - E\frac{\partial S}{\partial E} - N\frac{\partial S}{\partial N} + \cdots$$

but

$$\frac{\partial S}{\partial E} = \frac{1}{T}$$

$$\frac{\partial S}{\partial N} = -\frac{\mu}{T}$$

so that

$$S(E_T - E, N_T - N) = S(E_T, N_T) - \frac{E}{T} + \frac{\mu N}{T}$$

assuming we can ignore the higher terms. Then

$$P(E, N) \propto e^{S(E_T, N_T)/k} e^{-(E - \mu N)/kT}.$$

But $e^{S(E_T, N_T)/k}$ is simply a constant, so we finally obtain the probability

$$P(E, N) \propto e^{-(E - \mu N)/kT}. \tag{1.25}$$

This is the probability that a system held a temperature T and chemical potential μ will be found in a microstate of energy E, with N particles. The exponential factor $e^{-(E - \mu N)/kT}$ is sometimes known as the Gibbs factor, the Gibbs distribution function or the grand canonical distribution function.

1.4. Thermodynamic Averages

The importance of the previously derived probability distribution functions is that they may be used in calculating average (observed) values of various macroscopic properties of systems. In this way the aims of Statistical Mechanics, as outlined in Sec. 1.1.1 are achieved.

1.4.1. *The partition function*

The probability that a system is in the jth microstate, of energy $E_j(N, V)$ is given by the Boltzmann factor, which we write as:

$$P_j(N, V, T) = \frac{e^{-E_j(N, V)/kT}}{Z(N, V, T)} \tag{1.26}$$

where the normalization constant Z is given by

$$Z(N, V, T) = \sum_i e^{-E_i(N, V)/kT}. \tag{1.27}$$

Here we have been particular to indicate the functional dependencies. Energy eigenstates depend on the size of the system (standing waves), and the number of particles. And we are considering our system to be in thermal contact with a heat bath; thus the temperature dependence. We do not, however, allow particle interchange.

The quantity Z is called the (canonical) *partition function*. Although it has been introduced simply as a normalisation factor, we shall see that it is a very useful quantity indeed.

1.4.2. *Generalised expression for entropy*

For an isolated system the micro–macro connection is given by the Boltzmann formula $S = k \ln \Omega$, where Ω is a function of the extensive variables of the system

$$\Omega = \Omega(E, V, N).$$

But now, at a specified temperature, the energy E is not fixed, rather it fluctuates about a mean value $\langle E \rangle$.

To make the micro–macro connection when E is not fixed we must generalise the Boltzmann expression for entropy by looking at a collection of (macroscopically) identical systems in thermal contact. The composite system may be regarded as being isolated, so to *that* we may apply the rule $S = k \ln \Omega$, and then the mean entropy of a representative single system may be found.

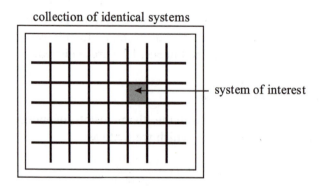

collection of identical systems

system of interest

Fig. 1.6.   Gibbs* ensemble for evaluating generalised entropy.

---

*The reason for calling this a *Gibbs* ensemble will become clear when you have encountered Sec. 1.6.2.

Let us consider M identical systems, and let there be n_j of these systems in the jth microstate. We assume that this is a collection of a very large number of systems. Then the systems other than that of particular interest may be regarded as a heat bath.

The number of possible microstates of the composite system corresponds to the number of ways of rearranging the subsystems:

$$\Omega = \frac{M!}{n_1! n_2! n_3! \cdots} \tag{1.28}$$

and the total entropy of the composite system is then

$$S_{\text{tot}} = k \ln \left(\frac{M!}{n_1! n_2! n_3! \cdots} \right). \tag{1.29}$$

Since all the numbers here are large, we may make use of Stirling's approximation for the logarithm of a factorial, $\ln n! \approx n \ln n - n$, so that

$$S_{\text{tot}} = k \left(M \ln M - \sum_j n_j \ln n_j \right).$$

Now if we express the first M as $\sum_j n_j$ then

$$S_{\text{tot}} = k \left(\sum_j n_j \ln M - \sum_j n_j \ln n_j \right)$$

$$= -k \sum_j n_j \ln \left(\frac{n_j}{M} \right).$$

We are interested in the mean entropy of our particular system. We have been considering a composite of M systems, so the mean entropy is simply the total entropy divided by M. Thus

$$S = -k \sum_j \frac{n_j}{M} \ln \left(\frac{n_j}{M} \right).$$

But n_j/M is the fraction of systems in the jth state, or the probability of finding our representative system in the jth state:

$$P_j = \frac{n_j}{M}.$$

So we can now express the entropy of a non-isolated system in terms of the state probabilities as

$$S = -k \sum_j P_j \ln P_j, \tag{1.30}$$

or

$$S = -k\langle \ln P_j \rangle, \tag{1.31}$$

the average value of the logarithm of the P_js. This is the Gibbs expression for the entropy. For an isolated system this reduces to the original Boltzmann expression.

1.4.3. *Free energy*

In the new expression for entropy we actually know the values for the probabilities — they are given by the Boltzmann factor:

$$P_j(N, V, T) = \frac{e^{-E_j(N,V)/kT}}{Z(N, V, T)}$$

where we recall that the normalisation factor is given by the sum over states, the partition function Z

$$Z(N, V, T) = \sum_i e^{-E_i(N,V)/kT}.$$

We then have

$$\ln P_j = -\left(\frac{E_j}{kT} + \ln Z\right).$$

Thus

$$S = k\left\langle \frac{E_j}{kT} + \ln Z \right\rangle$$

and since Z is independent of j we have

$$S = \frac{\langle E \rangle}{T} + k \ln Z.$$

Now in the spirit of thermodynamics we do not distinguish between mean and actual values — since fluctuations will be of order $1/\sqrt{N}$. Thus we write

$$E - TS = -kT \ln Z.$$

The quantity $E - TS$ is rather important and it is given a special name: *Helmholtz free energy*, or simply *free energy*. The symbol F is used.

$$F = E - TS \tag{1.32}$$

so that we can write

$$F = -kT \ln Z. \tag{1.33}$$

1.4.4. Thermodynamic variables

A host of thermodynamic variables can be obtained from the partition function. This is seen from the differential of the free energy. Since

$$dE = TdS - pdV + \mu dN$$

it follows that

$$dF = -SdT - pdV + \mu dN.$$

We can then identify the various partial derivatives:

$$\left.\begin{aligned}
S &= -\left.\frac{\partial F}{\partial T}\right|_{V,N} = kT \left.\frac{\partial \ln Z}{\partial T}\right|_{V,N} + k \ln Z \\[2mm]
p &= -\left.\frac{\partial F}{\partial V}\right|_{T,N} = kT \left.\frac{\partial \ln Z}{\partial V}\right|_{T,N} \\[2mm]
\mu &= \left.\frac{\partial F}{\partial N}\right|_{T,V} = -kT \left.\frac{\partial \ln Z}{\partial N}\right|_{T,V}
\end{aligned}\right\} \tag{1.34}$$

Since $E = F + TS$ we can then express the internal energy as

$$E = kT^2 \left.\frac{\partial \ln Z}{\partial T}\right|_{V,N}. \tag{1.35}$$

Thus we see that once the partition function is evaluated by summing over the states, all relevant thermodynamic variables can be obtained by differentiating Z.

1.4.5. Fluctuations

An isolated system has a well-defined energy. A system in contact with a heat bath has a well-defined temperature. However it is continually exchanging energy with the heat bath. We calculate the *average* value of the energy from

$$\langle E \rangle = kT^2 \frac{\partial \ln Z}{\partial T}\bigg|_{V,N}$$

but the *instantaneous* value of the energy in the system will be fluctuating about this value. What is the magnitude of these fluctuations? (In this section we denote the instantaneous energy by E and the average energy by $\langle E \rangle$.)

We shall evaluate the RMS (root mean square) of the energy fluctuations σ_E, defined by

$$\sigma_E = \langle (E - \langle E \rangle)^2 \rangle^{1/2}. \tag{1.36}$$

By expanding this out we obtain

$$\begin{aligned}
\sigma_E^2 &= \langle E^2 \rangle - 2\langle E \rangle^2 + \langle E \rangle^2 \\
&= \langle E^2 \rangle - \langle E \rangle^2. \tag{1.37}
\end{aligned}$$

We evaluate σ_E in the following way. Starting from the expression for $\langle E \rangle$ as

$$\langle E \rangle = \frac{1}{Z} \sum_j E_j e^{-E_j/kT}$$

we see that we could obtain $\langle E^2 \rangle$ by differentiating with respect to temperature so that another E_j comes down in the summation. It is simpler to obtain Z on the left-hand side first.

$$\langle E \rangle \sum_j e^{-E_j/kT} = \sum_j E_j e^{-E_j/kT}.$$

We differentiate this with respect to temperature (at constant volume):

$$\frac{\partial \langle E \rangle}{\partial T} \sum_j e^{-E_j/kT} + \frac{\langle E \rangle}{kT^2} \sum_j E_j e^{-E_j/kT} = \frac{1}{kT^2} \sum_j E_j^2 e^{-E_j/kT}.$$

This is then divided by the partition function, to give

$$\frac{\partial \langle E \rangle}{\partial T} + \frac{\langle E \rangle^2}{kT^2} = \frac{\langle E^2 \rangle}{kT^2},$$

or

$$\langle E^2 \rangle - \langle E \rangle^2 = kT^2 \frac{\partial \langle E \rangle}{\partial T}. \tag{1.38}$$

This may be written as

$$\langle E^2 \rangle - \langle E \rangle^2 = kT^2 C_V \tag{1.39}$$

since we recognise the derivative of energy, $\partial E / \partial T$ as the thermal capacity. Thus the RMS variation in the energy is given by

$$\sigma_E = \sqrt{kT^2 C_V}. \tag{1.40}$$

Since C_V and $\langle E \rangle$ are both proportional to the number of particles in the system, the *fractional* fluctuations in energy vary as

$$\frac{\sigma_E}{\langle E \rangle} \sim \frac{1}{\sqrt{N}} \tag{1.41}$$

which gets smaller and smaller as N increases.

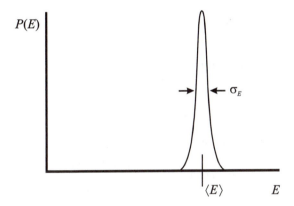

Fig. 1.7.   Fluctuations in energy for a system at fixed temperature.

We thus see that the importance of fluctuations vanishes in the thermodynamic limit $N \to \infty$, $V \to \infty$ while N/V remains constant. And it is in this limit that statistical mechanics has its greatest applicability.

1.4.6.   *The grand partition function*

Here we are concerned with systems of variable numbers of particles. The energy of a (many-body) state will depend on the number of particles in the system. As before, we label the (many-body) states of the system by j, but note that the jth state will be different for different N. In other words, we need the pair $\{N, j\}$ for specification of a state. The probability that a system has N particles and is in the jth microstate, corresponding to a total energy $E_{N,j}$ is given by the Gibbs factor, which we write as:

$$P_{N,j}(V, T, \mu) = \frac{e^{-[E_{N,j}(N,V)-\mu N]/kT}}{\Xi(V, T, \mu)} \tag{1.42}$$

where the normalization constant Ξ is given by

$$\Xi(V, T, \mu) = \sum_{N,j} e^{-[E_{N,j}(N,V)-\mu N]/kT}. \tag{1.43}$$

Here both T and μ are properties of the bath. The quantity Ξ is called the *grand partition function*. It may also be written in terms of the canonical partition function Z for different N:

$$\Xi(V, T, \mu) = \sum_{N} Z(N, V, T)e^{\mu N/kT}. \tag{1.44}$$

We will see that Ξ is also a useful quantity.

1.4.7. *The grand potential*

The generalised expression for entropy in this case is

$$S = -k\langle \ln P_{N,j} \rangle.$$

Here the probabilities are given by the Gibbs factor:

$$P_{N,j}(V, T, \mu) = \frac{e^{-[E_{N,j}(N,V)-\mu N]/kT}}{\Xi(V, T, \mu)}$$

where the normalisation factor, the sum over states, is the grand partition function Ξ. We then have

$$\ln P_{N,j} = -\left(\frac{E_{N,j}}{kT} - \frac{\mu N}{kT} + \ln \Xi \right).$$

Thus

$$S = k\left\langle \frac{E_{N,j}}{kT} - \frac{\mu N}{kT} + \ln \Xi \right\rangle$$

which is given by

$$S = \frac{\langle E \rangle}{T} - \frac{\mu \langle N \rangle}{T} + k \ln \Xi. \tag{1.45}$$

Now in the spirit of thermodynamics we do not distinguish between mean and actual values — since fluctuations will be of order $1/\sqrt{N}$. Thus we write

$$E - TS + \mu N = -kT \ln \Xi. \tag{1.46}$$

The quantity $E - TS + \mu N$ is equal to $-pV$ by the Gibbs–Duhem/Euler relation (see Appendix 1) so that we can write

$$pV = kT \ln \Xi. \tag{1.47}$$

The quantity pV is referred to as the *grand potential*.

1.4.8. *Thermodynamic variables*

Just as with the partition function, a host of thermodynamic variables can be obtained from the grand partition function. This is seen from the differential of the grand potential. Since

$$dE = TdS - pdV + \mu dN,$$

we can subtract Vdp from both sides to give

$$d(pV) = SdT + pdV + Nd\mu.$$

We can then identify the various partial derivatives:

$$\left.\begin{aligned}
S &= \left.\frac{\partial pV}{T}\right|_{V,\mu} = kT \left.\frac{\partial \ln \Xi}{\partial T}\right|_{V,\mu} + k \ln \Xi \\
p &= \left.\frac{\partial pV}{\partial V}\right|_{T,\mu} = kT \left.\frac{\partial \ln \Xi}{\partial V}\right|_{T,\mu} = \frac{kT}{V} \ln \Xi \\
N &= \left.\frac{\partial pV}{\partial \mu}\right|_{T,V} = kT \left.\frac{\partial \ln \Xi}{\partial \mu}\right|_{T,V}.
\end{aligned}\right\} \tag{1.48}$$

Thus we see that once the grand partition function is evaluated by summing over the states, all relevant thermodynamic variables can be obtained by differentiating Ξ.

1.5. Quantum Distributions

1.5.1. *Bosons and fermions*

All particles in nature can be classified into one of two groups according to the behaviour of their wave function under the exchange of identical particles. For simplicity let us consider just two identical particles. The wave function can then be represented as

$$\Psi = \Psi(r_1, r_2)$$

where

r_1 is the position of the first particle

and

r_2 is the position of the second particle.

Let us interchange the particles. We denote the operator that effects this by $\mathcal{P}$ (the permutation operator). Then

$$\mathcal{P}\Psi(r_1, r_2) = \Psi(r_2, r_1).$$

We are interested in the behaviour of the wave function under interchange of the particles. So far we have not drawn much of a conclusion. Let us now perform the swapping operation again. Then we have

$$\mathcal{P}^2\Psi(r_1, r_2) = \mathcal{P}\Psi(r_2, r_1) = \Psi(r_1, r_2);$$

the effect is to return the particles to their original states. Thus the operator $\mathcal{P}$ must obey

$$\mathcal{P}^2 = 1.$$

And taking the square root of this we find for $\mathcal{P}$

$$\mathcal{P} = \pm 1. \tag{1.49}$$

In other words the effect of swapping two identical particles is either to leave the wave function unchanged or to change the sign of the wave function.

This property continues for all time since the permutation operator commutes with the Hamiltonian. Thus all particles in nature belong to one class or the other. Particles for which

$\mathcal{P} = +1$ are called *bosons*

while those for which

$\mathcal{P} = -1$ are called *fermions*.

Fermions have the important property of not permitting multiple occupancy of quantum states. Consider two particles in the same state, at the same position r. The wave function is then

$$\Psi = \Psi(r, r).$$

Swapping over the particles we have

$$\mathcal{P}\Psi = -\Psi.$$

But $\Psi = \Psi(r, r)$ so that $\mathcal{P}\Psi = +\Psi$ since both particles are in the same state. The conclusion is that

$$\Psi(r, r) = -\Psi(r, r)$$

and this can only be so if

$$\Psi(r, r) = 0.$$

Now since Ψ is related to the *probability* of finding particles in the given state, the result $\Psi = 0$ implies a state of zero probability — an impossible state. We conclude that it is impossible to have more than one fermion in a given quantum state.

This discussion was carried out using r_1 and r_2 to denote *position* states. However that is not an important restriction. In fact, they could have designated any sort of quantum state and the same argument would follow. This is the explanation of the Pauli exclusion principle obeyed by electrons.

We conclude:

For bosons we can have any number of particles in a quantum state.

For fermions we can have either 0 or 1 particle in a quantum state.

But what determines whether a given particle is a boson or a fermion? The answer is provided by quantum field theory. And it depends on the *spin* of the particle. Particles whose spin angular momentum is an integral multiple of $\hbar$ are bosons while particles whose spin angular momentum is integer plus a half $\hbar$ are fermions. (In quantum theory $\hbar/2$ is the smallest unit of spin angular momentum.) It is not straightforward to demonstrate this fundamental connection between spin and statistics. Feynman's heroic attempt is contained in his 1986 Dirac memorial lecture.[12] However a slightly more accessible account is contained in Tomonaga's book *The Story of Spin*.[13]

For some elementary particles we have:

$$\left.\begin{array}{c}\text{electrons}\\\text{protons}\\\text{neutrons}\end{array}\right\}S=\frac{1}{2}\;\rightarrow\;\text{fermions}$$

$$\left.\begin{array}{c}\text{photons}\quad S=1\\\left.\begin{array}{c}\pi\text{ mesons}\\\kappa\text{ mesons}\end{array}\right\}S=0\end{array}\right\}\rightarrow\text{bosons.}$$

For composite particles (such as atoms) we simply add the spins of the constituent parts. And since protons, neutrons and electrons are all fermions we can say:

$$\text{Odd number of fermions}\rightarrow\text{fermion}$$
$$\text{Even number of fermions}\rightarrow\text{boson.}$$

The classic example of this is the two isotopes of helium. Thus

$$^3\text{He is a fermion}$$
$$^4\text{He is a boson.}$$

At low temperatures these isotopes have very different behaviour.

1.5.2. *Grand potential for identical particles*

The grand potential allows the treatment of systems of variable numbers of particles. We may exploit this in the study of systems of non-interacting (or weakly-interacting) particles in the following way. We focus attention on a single-particle state, which we label by k. The state of the entire system is specified when we know how many particles are in each different (single-particle) quantum state.

$$\text{many-particle state}\equiv\{n_1,n_2,\ldots,n_k,\ldots\}$$
$$\text{Energy of state}\equiv\sum_k n_k\varepsilon_k$$
$$\text{No. of particles}\equiv\sum_k n_k.$$

Here ε_k is the energy of the kth single-particle state. Note that the ε_k are independent of N.

Now since the formalism of the grand potential is appropriate for systems that exchange particles and energy with their surroundings, we may now consider as our "system" the subsystem comprising the particles in a given state k. For this subsystem

$$E = n_k \varepsilon_k$$
$$N = n_k$$

so that the probability of observing this, i.e. the probability of finding n_k particles in the kth state (provided this is allowed by the statistics) is

$$P_{n_k}(V, T, \mu) = \frac{e^{-(n_k \varepsilon_k - n_k \mu)/kT}}{\Xi_k}$$

where the grand partition function for the subsystem can be written

$$\Xi_k = \sum_{n_k} \left\{ e^{-(\varepsilon_k - \mu)/kT} \right\}^{n_k}.$$

Here n_k takes only values 0 and 1 for fermions and $0, 1, 2, \ldots, \infty$ for bosons. The grand potential for the "system" is

$$(pV)_k = kT \ln \Xi_k$$
$$= kT \ln \sum_{n_k} \left\{ e^{-(\varepsilon_k - \mu)/kT} \right\}^{n_k}. \tag{1.50}$$

The grand partition function for the entire system is the product

$$\Xi = \prod_k \Xi_k$$

so that the grand potential (and any other extensive quantity) for the entire system is found by summing over all single-particle state contributions:

$$pV = \sum_k (pV)_k.$$

1.5.3. *The Fermi distribution*

For fermions the grand potential for the single state is

$$(pV)_k = kT \ln \sum_{n_k = 0,1} \left\{ e^{-(\varepsilon_k - \mu)/kT} \right\}^{n_k}$$
$$= kT \ln \left\{ 1 + e^{-(\varepsilon_k - \mu)/kT} \right\}. \tag{1.51}$$

From this we can find the mean number of particles in the state using

$$\bar{n}_k = \left. \frac{\partial (pV)_k}{\partial \mu} \right|_{T,V} = \frac{e^{-(\varepsilon_k - \mu)/kT}}{1 + e^{-(\varepsilon_k - \mu)/kT}}$$
$$= \frac{1}{e^{(\varepsilon_k - \mu)/kT} + 1}. \tag{1.52}$$

This is known as the Fermi–Dirac distribution function.

The grand potential for the entire system of fermions is found by summing the single-state grand potentials

$$pV = kT \sum_k \ln \left\{ 1 + e^{-(\varepsilon_k - \mu)/kT} \right\}. \tag{1.53}$$

1.5.4. The Bose distribution

For bosons the grand potential for the single state is

$$
\begin{aligned}
(pV)_k &= kT \ln \sum_{n_k=0}^{\infty} \left\{ e^{-(\varepsilon_k - \mu)/kT} \right\}^{n_k} \\
&= kT \ln \left\{ \frac{1}{1 - e^{-(\varepsilon_k - \mu)/kT}} \right\} \\
&= -kT \ln \left\{ 1 - e^{-(\varepsilon_k - \mu)/kT} \right\}
\end{aligned} \tag{1.54}
$$

assuming the geometric progression is convergent. From this we can find the mean number of particles in the state using

$$
\begin{aligned}
\bar{n}_k &= \left. \frac{\partial (pV)_k}{\partial \mu} \right|_{T,V} = \frac{e^{-(\varepsilon_k - \mu)/kT}}{1 - e^{-(\varepsilon_k - \mu)/kT}} \\
&= \frac{1}{e^{(\varepsilon_k - \mu)/kT} - 1}.
\end{aligned} \tag{1.55}
$$

This is known as the Bose–Einstein distribution function.

The grand potential for the entire system of bosons is found by summing the single-state grand potentials

$$pV = -kT \sum_k \ln \left\{ 1 - e^{-(\varepsilon_k - \mu)/kT} \right\}. \tag{1.56}$$

(An elegant derivation of the Bose and the Fermi distributions which indicates how the $+$ and $-$ sign in the denominators arises directly from the eigenvalue of the $\mathcal{P}$ operator is given in the Quantum Mechanics text book by Merzbacher.[14] Beware, however — it uses the method of Second Quantisation.)

1.5.5. The classical limit — The Maxwell distribution

The Bose–Einstein and Fermi–Dirac distributions give the mean numbers of particles in the microstate of energy ε_j as a function of $(\varepsilon_j - \mu)/kT$. When this quantity is large we observe two things. Firstly, the denominator of the distributions will be very much larger than one, so the $+1$ or -1 distinguishing fermions from bosons may be neglected. And secondly, the large value for the denominator means that $\bar{n}_j$, the mean occupation of the state, will be very much less than unity.

This condition will apply to all states, down to the ground state of $\varepsilon_j = 0$, if μ/kT is large and negative. This is the classical limit where the issue of multiple state occupancy does not arise and the distinction between fermions and bosons becomes unimportant. We refer to such (imaginary) particle as maxwellons, obeying Maxwell–Boltzmann statistics. Thus, for these particles the mean number of particles in the state is given by

$$\bar{n}_k = e^{-(\varepsilon_k - \mu)/kT}.$$

The three distribution functions are shown in Fig. 1.8. Observe, in particular, that when $\mu = \varepsilon$ the Fermi occupation is one half, the Maxwell occupation is unity and the Bose occupation is infinite.

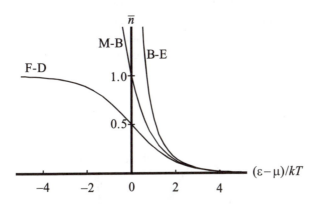

Fig. 1.8. Fermi–Dirac, Maxwell–Boltzmann and Bose–Einstein distribution functions.

1.6. Classical Statistical Mechanics

1.6.1. *Phase space and classical states*

The formalism of statistical mechanics developed thus far relies very much, at the microscopic level, on the use of (micro)states. We count the number of states, we sum over states, etc. This is all very convenient to do within the framework of a quantum-mechanical description of systems where states of a (finite or bound) system are discrete, but what about classical systems. How is the formalism of *classical* statistical mechanics developed — what is a "classical state"?

A classical microstate is a state for which we have complete information at the microscopic level. In other words, we must know the position and velocity of all particles in the system. (Position *and* velocity, since the equations of motion — Newton's laws — are second-order differential equations in time.)

For reasons related to the Lagrangian and the Hamiltonian formulation of mechanics it proves convenient to work in terms of the position and the *momentum* rather than velocity of the particles in the system. This is because it is then possible to work in terms of *generalised* coordinates and momenta — such as angles and angular momenta — in a completely independent way; one is not constrained to a particular coordinate system. Thus we will say that a classical microstate is specified by the coordinates and the momenta of all the constituent particles. A single particle has three coordinates x, y, z and three momentum components p_x, p_y, p_z so it needs *six* components to specify its state. The coordinates are conventionally denoted by q and the momenta by p. This pq space is called *phase space*.

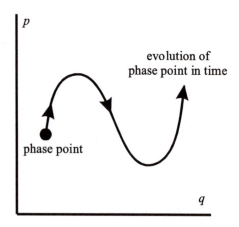

Fig. 1.9.   Phase space.

The state of a particle is denoted by a *point* in phase space. Its evolution in time is represented by a *curve* in phase space. It should be evident that during its evolution the phase curve of a point cannot intersect since there is a unique path proceeding from each location in phase space.

A quantum state may be specified by a set of quantum numbers. A classical system is specified when we know the position and momentum coordinates of all the particles. But what is the classical analogue of the quantum state? We need to know this so that we may use the analogous Boltzmann expression for entropy, and we want to understand how to construct partition functions for classical systems. And fundamentally, of course, the definition of the classical state must be consistent with the Fundamental Postulate.

The classical analogue of a quantum state, a "classical state" is specified as a *cell* in phase space. For a single particle with coordinates q_x, q_y, q_z and momentum components p_x, p_y, p_z we have

$$\text{"classical state"} \propto \Delta q_x \Delta q_y \Delta q_z \Delta p_x \Delta p_y \Delta p_z = \Delta^3 p \Delta^3 q \text{ for simplicity.}$$

This will not be proved, but it is eminently plausible. The question of the constant of proportionality is interesting. The value of this constant may be ascertained by comparing the partition functions as calculated classically and quantum-mechanically in the classical limit (high temperature/low density). One finds that for each pq pair there is a factor of h^{-1}. This is certainly dimensionally correct, but what business has Planck's constant to appear in a fundamentally classical result? For a single particle we then have

$$\text{"classical state"} \text{ corresponds to } \Delta q_x \Delta q_y \Delta q_z \Delta p_x \Delta p_y \Delta p_z = h^3.$$

The number of "classical states" in a small region of phase space is then $\Delta^3 p \Delta^3 q / h$. The general rule, then, is that the sums over states in the quantum case correspond to integrals over phase space in the classical case:

$$\sum_{\substack{\text{single} \\ \text{particle} \\ \text{states}}} \rightarrow \frac{1}{h^3} \int d^3 p \, d^3 q. \tag{1.57}$$

1.6.2. *Boltzmann and Gibbs phase spaces*

The microstate of a system is represented by a point in phase space. A different microstate will be represented by a different point in phase space. In developing the statistical approach to mechanics we must talk about different microstates — so we are considering different points in phase space and probabilities associated with them. Boltzmann and Gibbs looked at this in different ways. Boltzmann's idea was that a gas of N particles would be represented by N points in the six-dimensional phase space. The evolution of the state of the system with time is then described by the "flow" of the "gas" of points in the phase space. This sort of argument only works for weakly interacting particles, since later arguments are based on the movement of the points in phase space being independent.

Gibbs adopted a rather more general approach. He regarded the state of a system of N particles as being specified by a single point in a $6N$-dimensional phase space. The six-dimensional phase space of Boltzmann is sometimes referred to as μ-space and the $6N$-dimensional phase space of Gibbs is sometimes referred to as Γ-space.

In both cases, one applies probabilistic arguments to the collection of points in phase space. This collection is called an *ensemble*. So in Boltzmann's view a single particle is the system and the N particles comprise the ensemble while in Gibbs's view the assembly of particles is the system and many imaginary copies of the system comprise the ensemble. In the Boltzmann case one performs averages over the possible states of a single particle, while in the Gibbs case one is considering possible states of the entire system and applying probabilistic arguments to those.

Both views are useful. The Boltzmann approach is easier to picture but it can only be applied to weakly interacting particles. The Gibbs approach is more powerful as it can be applied to strongly interacting systems where the particles cannot be regarded as being even *approximately* independent.

The probability of finding a system in a microstate in the region $dp\,dq$ of phase space is given by $\rho\,dp\,dq$ where ρ is the density of representative points in the phase space. So the probability density of the microstate p, q is given by $\rho(p, q)$.

1.6.3. *The Fundamental Postulate in the classical case*

If we say that all (quantum) states are equally likely, then the classical analogue will be that all points in phase space are equally likely. The quantum version of the Fundamental Postulate refers to an *isolated system*. This is a system for which E, V and N are fixed. Classically when the energy is fixed, this restricts the accessible region of phase space to a constant-energy hypersurface. Similarly, fixing V and N determines which regions of phase space are available. The classical version of the Fundamental

Postulate then states that **for an isolated system all available regions of phase space on the constant energy hypersurface are equally likely.**

The probability of a macrostate corresponding to a region of phase space will then be proportional to the number of phase points in the region. So correctly normalised, it will be given by the *density* of points $\rho(p, q)$ where by p and q we mean the set of all momentum coordinates and all position coordinates.

1.6.4. *The classical partition function*

The classical analogue of the quantum partition function is given by

$$Z = \frac{1}{h^{3N}} \int e^{-H(p_i, q_i)/kT} \mathrm{d}^{3N}p \, \mathrm{d}^{3N}q \qquad (1.58)$$

when we are considering distinguishable particles. For indistinguishable particles there is the factor of $1/N!$ (Sec. 2.3.4)

$$Z = \frac{1}{N!h^{3N}} \int e^{-H(p_i, q_i)/kT} \mathrm{d}^{3N}p \, \mathrm{d}^{3N}q \qquad (1.59)$$

and there is the possibility of factorising the expression for Z.

The function $H(p_i, q_i)$ is the energy of the system expressed as a function of the position and momentum coordinates q_i and p_i.

1.6.5. *The equipartition theorem*

The equipartition theorem is concerned with the internal energy associated with individual degrees of freedom of a system. It has important consequences for the behaviour of the thermal capacity of classical systems.

We ask the question "What is the internal energy associated with a given degree of freedom — say p_i?". That is easy to write down:

$$\langle E_i \rangle = \frac{1}{Z} \int E_i e^{-E(q_1 \cdots q_i \cdots q_N, p_1 \cdots p_i \cdots p_N)/kT} \mathrm{d}^{3N}q \, \mathrm{d}^{3N}p. \qquad (1.60)$$

Assuming the energy E_i depends only on the p_i and not on the other p's and q's, we can factorise that bit out of the exponential and write the integral as

$$\langle E_i \rangle = \frac{\int E_i e^{-E_i/kT} \mathrm{d}p_i \overset{\lceil \text{no } p_i \rceil}{\int e^{-E(q_1 \cdots q_i \cdots q_N, p_1 \cdots p_N)/kT} \mathrm{d}^{3N}q \, \mathrm{d}^{3N-1}p}}{\int e^{-E_i/kT} \mathrm{d}p_i \int [\text{same integral as above} - \text{no } p_i]}.$$

So the second integral in numerator and denominator cancel, leaving the simple expression

$$\langle E_i \rangle = \frac{\int E_i e^{-E_i/kT} dp_i}{\int e^{-E_i/kT} dp_i}.$$

This may be simplified by writing $\beta = 1/kT$ and using

$$E_i e^{-\beta E_i} = -\frac{\partial}{\partial \beta} e^{-\beta E_i}$$

so that

$$\langle E_i \rangle = -\frac{\frac{\partial}{\partial \beta} \int e^{-\beta E_i} dp_i}{\int e^{-\beta E_i} dp_i}$$

or

$$\langle E_i \rangle = -\frac{\partial}{\partial \beta} \ln \int e^{-\beta E_i} dp_i.$$

At this stage we must be more specific about the functional form of $E_i(p_i)$. Since p_i is a momentum then for a classical particle $E_i(p_i) = p_i^2/2m$ — a quadratic dependence. For simplicity let us write simply

$$E_i = bp_i^2$$

for some constant b. The integral is then

$$\int e^{-\beta E_i} dp_i = \int e^{-\beta b p_i^2} dp_i.$$

We do not actually need to evaluate this. Remember that we are going to differentiate the logarithm of the integral with respect to β; all we want is the β-dependence. Let us make a change of variable and put

$$\beta p_i^2 = y^2.$$

The integral then becomes

$$\beta^{-1/2} \int e^{-by^2} dy$$

so that

$$\langle E_i \rangle = -\frac{\partial}{\partial \beta} \ln \left(\beta^{-1/2} \int e^{-by^2} dy \right)$$

$$= -\frac{\partial}{\partial \beta} \left\{ -\frac{1}{2} \ln \beta + \ln \int e^{-by^2} dy \right\}.$$

The second term is independent of β so upon differentiation it vanishes. Thus differentiating we obtain

$$\langle E_i \rangle = \frac{1}{2\beta}$$

or, in terms of T:

$$\langle E_i \rangle = \frac{1}{2}kT. \tag{1.61}$$

The general conclusion here may be stated as the **Equipartition theorem**: For a classical (non-quantum) system each degree of freedom with a quadratic dependence on coordinate or momentum gives a contribution to the internal energy of $kT/2$.

[Incidentally, if $E_i \propto q_i^n$ or p_i^n then $\langle E_i \rangle = kT/n$.]

1.6.6. *Consequences of equipartition*

We consider two examples — lattice vibrations, and a gas of particles.

For the case of lattice vibrations each atom is essentially three harmonic oscillators, one in the x, y and z directions. Thus for N atoms we have $3N$ harmonic oscillators. Now in this case, *both* the position and momentum coordinates contribute a quadratic term to the energy. The internal energy is then

$$E = 3NkT \tag{1.62}$$

in the non-quantum (high temperature) limit. Differentiation with respect to temperature gives the isochoric thermal capacity

$$C_v = 3Nk$$
$$= 3R \text{ per mole.} \tag{1.63}$$

Considering now a gas of noninteracting particles, there is no contribution to the energy from the position coordinates. Only the momentum coordinates contribute a quadratic term to the energy and the internal energy is then

$$E = \frac{3}{2}NkT \tag{1.64}$$

in the non-quantum (high temperature) limit. Differentiation with respect to temperature gives the (isochoric) thermal capacity

$$C_v = \frac{3}{2}Nk$$

$$= \frac{3}{2}R \text{ per mole.} \tag{1.65}$$

The thermal capacity of the solid is double that of the fluid because only in the solid do the position coordinates contribute to the internal energy.

[In fact the walls of a box of gas can be modelled as an oscillator with a power law potential $V \propto x^n$ where $n \to \infty$.]

Equipartition breaks down when quantum effects become important. In Sec. 2.3.3 we shall see that the internal energy of a *single* quantum free particle corresponds to $\frac{3}{2}kT$: the equipartition value for the three spatial degrees of freedom. However, once we have a collection of N identical particles comprising a quantum gas, the internal energy is given, in Sec. 2.4.5, by

$$E = \frac{3}{2}NkT\left\{1 + a\sqrt{\frac{2}{\pi}\frac{1}{6}}\left(\frac{\varepsilon_F}{kT}\right)^{3/2} + a^2 \cdots\right\},$$

where $a = +1$ for fermions, zero for "classical" particles and -1 for bosons. The equipartition result occurs at high temperatures and as the gas cools, quantum effects become important. For fermions the internal energy increases above the equipartition value, while for bosons the internal energy decreases below the equipartition value.

In Problem 1.16 you will see that the internal energy of a quantum harmonic oscillator may be written as

$$E = kT + \frac{\hbar^2\omega^2}{12kT} + \cdots.$$

The first term represents the high-temperature equipartition value. The second (and higher) terms indicate the internal energy increasing above its high-temperature value as the temperature is lowered.

1.6.7. *Liouville's theorem*

We ask the question "How does a macroscopic system evolve in time?". The answer is that it will develop in accordance with the Second Law of thermodynamics; the system evolves to the state of maximum entropy consistent with the constraints imposed. Can this be understood from microscopic first principles? In other words, can the law of entropy increase be derived from Newton's laws? Both Boltzmann and Gibbs agonised over this.

We need a definition of entropy which will be suitable for use in the classical case. The problem is that there are no *discrete* states now, since p and q can vary continuously. By analogy with the generalised expression for entropy:

$$S = -k \sum_j P_j \ln P_j,$$

since the probability (density) is given by the density of points in phase space, we now have

$$S = -k \int \rho \ln \rho \, dp dq. \tag{1.66}$$

It is essentially (minus) the average of the logarithm of the density of points in phase space.

If one calculates the way points move around phase space under the influence of the laws of mechanics one finds that the "flow" is incompressible. Thus the density remains constant. This result is known as Liouville's theorem. We need the machinery of Hamiltonian mechanics to show this. If you are happy with Hamiltonian mechanics the proof is sketched below. But the implication is that since ρ remains constant then the entropy remains constant, so the Second Law of thermodynamics seems to be inconsistent with the laws of mechanics at the microscopic level.

To demonstrate Liouville's theorem we first note that the flow of points in phase space must obey the equation of continuity, since the number of points is conserved:

$$\frac{\partial \rho}{\partial t} + \text{div } \mathbf{v}\rho = 0. \tag{1.67}$$

However, in this case ρ depends on the position and momentum coordinates, q and p. Thus the divergence contains all the $\partial/\partial q$ derivatives and all the $\partial/\partial p$ derivatives. And the "velocity" $\mathbf{v}$ has components dp/dt as well as the usual dq/dt. Thus the divergence term is actually

$$\text{div } \mathbf{v}\rho = \frac{\partial}{\partial p}\left(\frac{dp}{dt}\rho\right) + \frac{\partial}{\partial q}\left(\frac{dq}{dt}\rho\right)$$

(these equations really contain all the q and p coordinates; the above, as elsewhere, is a shorthand simplification). We expand the p and q derivatives to give

$$\text{div}\,\mathbf{v}\rho = \left(\frac{\partial}{\partial p}\frac{dp}{dt} + \frac{\partial}{\partial q}\frac{dq}{dt}\right)\rho + \frac{\partial\rho}{\partial p}\frac{dp}{dt} + \frac{\partial\rho}{\partial q}\frac{dq}{dt}$$

and then we use Hamilton's equations

$$\frac{dp}{dt} = -\frac{\partial H}{\partial q}, \quad \frac{dq}{dt} = \frac{\partial H}{\partial p}$$

in the first bracket. Then

$$\frac{\partial}{\partial p}\frac{dp}{dt} + \frac{\partial}{\partial q}\frac{dq}{dt} = -\frac{\partial^2 H}{\partial p\partial q} + \frac{\partial^2 H}{\partial q\partial p} = 0$$

Thus we find that

$$\frac{\partial\rho}{\partial t} + \frac{\partial\rho}{\partial p}\frac{dp}{dt} + \frac{\partial\rho}{\partial q}\frac{dq}{dt} = 0.$$

But we recognise this as the *total* derivative of ρ; it is the derivative of the density when moving with the flow in phase space. This is zero. Thus as the representative points evolve and flow in phase space, the local density remains constant. This is the content of Liouville's theorem, expressed in its usual form as

$$\frac{d\rho}{dt} = 0. \tag{1.68}$$

We have a paradox: since ρ remains constant during evolution then the entropy remains constant; the Second Law of thermodynamics seems to be inconsistent with the laws of mechanics at the microscopic level.

1.6.8. *Boltzmann's H theorem*

The resolution of the paradox of the incompatibility between Liouville's theorem and the Second Law may be understood from the *nature* of the "flow" of points in phase space. Boltzmann defined a quantity H, which was the integral of $\rho \ln \rho$ over phase space and he obtained an equation of motion for H as a probabilistic differential equation for the flow of points into and out of regions of phase space. We shall adopt a variant of the approach of Gibbs to study the evolution of Boltzmann's H. Please note this H is neither the Hamiltonian of the previous section nor the enthalpy function; this is Boltzmann's H.

The flow of points in phase space is complicated. Since we have a given number of elements in our ensemble, the number of points in phase space is fixed. So Liouville's theorem is saying that the multidimensional volume occupied by the points is constant. But the flow can be "dendritic" with fingers spreading and splitting in all directions.

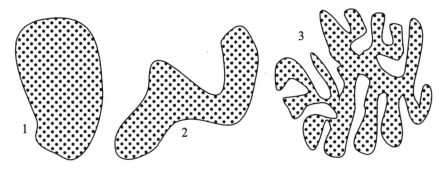

Fig. 1.10. Evolution of a region of phase space.

And as this happens, there will come a time when it is difficult to distinguish between what is an occupied region and what is an unoccupied region of phase space; they will be continually folded into each other. Gibbs argued that there was a scale in phase space, beyond which it was not possible (or at least reasonable) to discern. If the details of the state "3" above are too fine to discern, then it will simply appear as a region of greater volume and less density.

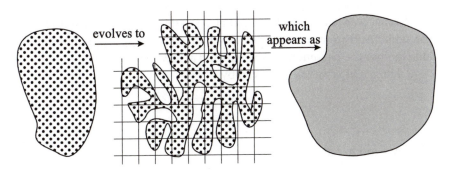

Fig. 1.11. Apparent reduction in density in phase space.

The procedure of taking an average over small regions of phase space is known as "coarse graining". Gibbs showed rigorously (rather than by just using diagrams as here) that the coarse grained density of points in phase space decreased as time proceeded.[15]

Thus the conclusion is that the coarse-grained H decreases, or the coarse-grained entropy increases.

There is a relation with quantum mechanics, which may be invoked in the question of coarse-graining. The volume of a "cell" in phase space is a product of p, q pairs. Now the Uncertainty Principle tells us that we cannot locate a point within a pq area better than Planck's constant. This gives the ultimate resolution that is achievable in specifying the state of a system — so at the fundamental level there is a firm justification for coarse-graining.

Quantum mechanics has a habit of popping up in the most unexpected areas of statistical thermodynamics. This theme continues into the next section.

1.7.   The Third Law of Thermodynamics

1.7.1.   *History of the Third Law*

The Third Law of thermodynamics arose as the result of experimental work in chemistry, principally by the famous chemist Nernst. He published what he called his "heat theorem" in 1906. A readable account of the history of the Third Law and the controversies surrounding its acceptance is given by Dugdale.[16]

Nernst measured the change in Gibbs free energy and the change in enthalpy for chemical reactions which started and finished at the same temperature. At lower and lower temperatures he found that the changes in G and the changes in H became closer and closer.

Nernst was led to conclude that at $T = 0$ the changes in G and H were the same. And from some elementary thermodynamic arguments he was able to infer the behaviour of the entropy at low temperatures.

Changes in H and G are given by

$$\Delta H = T\Delta S + V\Delta p$$
$$\Delta G = -S\Delta T + V\Delta p.$$

Thus ΔG and ΔH are related by

$$\Delta G = \Delta H - T\Delta S - S\Delta T$$

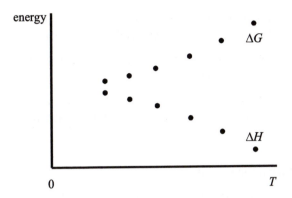

Fig. 1.12. Nernst's observations.

and if the temperature is the same before and after, $\Delta T = 0$, so then

$$\Delta G = \Delta H - T\Delta S.$$

This is a very important equation for chemists.

Now Nernst's observation may be stated as

$$\Delta H - \Delta G \to 0 \quad \text{as } T \to 0,$$

which implied that

$$T\Delta S \to 0 \quad \text{as } T \to 0.$$

1.7.2. *Entropy*

On the face of it this result is no surprise since the factor T will ensure the product $T\Delta S$ goes to zero. But Nernst took the result further. He studied *how fast* $\Delta H - \Delta G$ tended to zero. And his observation was that it always went faster than linearly. In other words he concluded that

$$\frac{\Delta H - \Delta G}{T} \to 0 \quad \text{as } T \to 0. \tag{1.69}$$

So even though $1/T$ was getting bigger and bigger, the quotient $(\Delta H - \Delta G)/T$ still tended to zero. But we know that

$$\frac{\Delta H - \Delta G}{T} = \Delta S.$$

So from this, Nernst drew the conclusion

$$\Delta S \to 0 \quad \text{as } T \to 0. \tag{1.70}$$

The entropy change in a process tends to zero at $T = 0$. The entropy thus remains a constant in any process at absolute zero. We conclude:

- The entropy of a body at zero temperature is a constant, independent of all other external parameters.

This was the conclusion of Nernst; it is sometimes called Nernst's heat theorem. It was subsequently to be developed into the Third Law of thermodynamics.

1.7.3. *Quantum viewpoint*

From the purely macroscopic perspective the Third Law is as stated above: at $T = 0$ the entropy of a body is a constant. And many conclusions can be drawn from this. One might ask the question "what is the constant?". However we do know that thermodynamic conclusions about measurable quantities are not influenced by any such additive constants since one usually differentiates to find observables. (But a constant of minus infinity, as found for the entropy of the classical ideal gas, might be problematic.)

If we want to ask about the constant then we must look into the microscopic model for the system under investigation. Recall the Boltzmann expression for entropy:

$$S = k \ln \Omega$$

where Ω is the number of microstates in the macrostate. Now consider the situation at $T = 0$. Then we know the system will be in its ground state, the lowest energy state. But this is a *unique* quantum state. Thus for the ground state

$$\Omega = 1$$

and so

$$S = 0.$$

Nernst's constant is thus zero and we then have the expression for the Third Law:

- As the absolute zero of temperature is approached the entropy of all bodies tends to zero.

We note that this applies specifically to bodies that are in *thermal equilibrium*. The Third Law can be summarised as

$$\frac{\partial S}{\partial \text{ anything}} \to 0 \quad \text{as } T \to 0. \tag{1.71}$$

The above discussion is actually an over-simplification. In reality there may be degeneracy in the ground state of the system; then the above argument appears to break down. However, recall that entropy is an extensive quantity and that the entropy of the system should be considered in the thermodynamic limit. In other words, strictly, we should examine how the intensive quantity S/V or S/N behaves in the limit $V \to \infty$, $N \to \infty$. If the degeneracy of the ground state is g then we must look at the behaviour of $\ln(g)/N$. This will tend to zero in the thermodynamic limit so long as g increases with N no faster than exponentially. This is the fundamental quantum-mechanical principle behind the Third Law. The interested reader should consult the paper by Leggett[17] for a deeper discussion of these points.

To complete this discussion, it is instructive to see how the Third Law would fail if classical mechanics were to apply down to $T = 0$. We saw, in Sec. 1.6.7, that the Gibbs expression for entropy:

$$S = -k \sum_j P_j \ln P_j$$

must be replaced, in the classical case by:

$$S = -k \int \rho \ln \rho \, dp \, dq,$$

where ρ is the density of points in phase space. This is necessary because in the classical case there are no discrete states and the momenta and coordinates, p and q, can vary continuously.

As the temperature is lowered, the mean energy of the system will decrease. And corresponding to this, the "volume" of phase space occupied will decrease. In particular, the momentum coordinates q will vary over a smaller and smaller range. In the $T \to 0$ limit the momentum range will become localised closer and closer to $q = 0$. The volume of occupied phase space shrinks to zero and the entropy thus tends to $-\infty$. This indeed is the limiting value indicated by the classical treatment of the Ideal Gas, in Sec. 2.3.3.

The Uncertainly Principle of quantum mechanics limits the low-temperature position–momentum of a system; you cannot localise points in phase space to a volume smaller than the appropriate power of Planck's constant. This fundamental limitation of the density of phase points recovers the Third Law. Thus again we see the intimate connection between quantum mechanics and the Third law.

The Second Law tells us that there is an absolute zero of temperature. Now we see that the Third Law tells us there is an absolute zero of entropy.

1.7.4. *Unattainability of absolute zero*

The Third Law has important implications concerning the possibility of cooling a body to absolute zero. Let us consider a sequence of adiabatic and isothermal operations on two systems, one obeying the Third Law and one not.

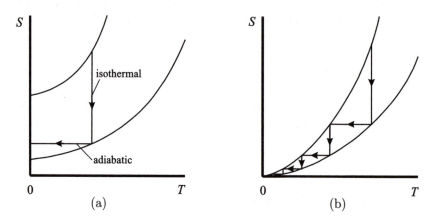

Fig. 1.13. Approaching absolute zero. System (a) not obeying the Third Law can get to $T = 0$ in two steps. System (b) obeying the Third Law cannot get to $T = 0$ in a finite number of steps.

Taking a sequence of adiabatics and isothermals between two values of some external parameter we see that the existence of the Third Law implies that you cannot get to $T = 0$ in a finite number of steps. This is, in fact, another possible statement of the Third Law.

Although one cannot get all the way to $T = 0$, it is possible to get closer and closer. Figure 1.14, adapted and extended from Pobel's book,[18] indicates the success in this venture.

1.7.5. *Heat capacity at low temperatures*

The Third Law has important consequences for the heat capacity of bodies at low temperatures. Since

$$C = \frac{\partial Q}{\partial T}$$

$$= T\frac{\partial S}{\partial T},$$ (1.72)

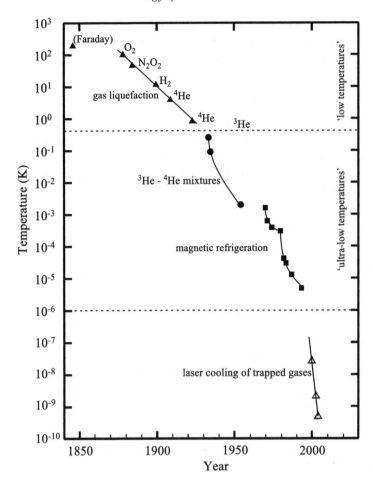

Fig. 1.14. The road to absolute zero.

and the Third Law tells us that

$$\frac{\partial S}{\partial T} \to 0 \quad \text{as } T \to 0,$$

we then have

$$C \to 0 \quad \text{as } T \to 0. \tag{1.73}$$

Classical models often give a constant heat capacity. For an ideal gas (Sec. 2.3.3)

$$C_V = \frac{3}{2} Nk$$

independent of temperature. The Third Law tells us that this cannot hold at low temperatures. And indeed we shall see that for both Fermi and Bose gases C_V does indeed go to zero at $T = 0$.

1.7.6. *Other consequences of the Third Law*

Most "response functions" or susceptibilities — generalised spring constants — go to zero or to a constant as $T \to 0$ as a consequence of the Third Law. This is best seen by examining the relevant Maxwell relation (Appendix A.2.6). For example consider the thermal expansion coefficient. The Maxwell relation here is

$$\left.\frac{\partial V}{\partial T}\right|_p = -\left.\frac{\partial S}{\partial p}\right|_T.$$

The right-hand side is zero by virtue of the Third Law. Thus we conclude that

$$\left.\frac{\partial V}{\partial T}\right|_p \to 0 \quad \text{as } T \to 0;$$

the expansion coefficient goes to zero.

An interesting example is the susceptibility of a paramagnet, which we will treat in Sec. 2.7. The connection with the model pV system is made *via*

$$M \to p$$
$$B \to V,$$

where M is the total magnetic moment. The magnetic susceptibility is (neglecting factors of μ_0)

$$\chi = \frac{1}{V}\frac{\partial M}{\partial B}$$

so that

$$\chi V \to \frac{\partial p}{\partial V}.$$

There is no Maxwell relation for this, but consider the variation of the susceptibility with temperature:

$$\frac{\partial(\chi V)}{\partial T} = \frac{\partial^2 M}{\partial T \partial B}$$

$$\to \frac{\partial^2 p}{\partial T \partial V}.$$

The order of differentiation can be reversed here. In other words

$$\frac{\partial}{\partial B}\frac{\partial M}{\partial T} \rightarrow \frac{\partial}{\partial V}\frac{\partial p}{\partial T}.$$

And now we do have a Maxwell relation:

$$\left.\frac{\partial p}{\partial T}\right|_V = \left.\frac{\partial S}{\partial V}\right|_T \quad \leftarrow \quad \left.\frac{\partial M}{\partial T}\right|_V = \left.\frac{\partial S}{\partial B}\right|_T.$$

The Third Law tells us that the right-hand side of these equations goes to zero as $T \rightarrow 0$. We conclude then that

$$\frac{\partial \chi}{\partial T} \rightarrow 0 \quad \text{as } T \rightarrow 0$$

or

$$\chi \rightarrow \text{const.} \quad \text{as } T \rightarrow 0. \tag{1.74}$$

The Third Law tells us that the magnetic susceptibility becomes constant as $T \rightarrow 0$. But what does Curie's law, Eqs. (2.76) and (2.77) say? This states

$$\chi = \frac{C}{T}$$

where C is the Curie constant. From this we conclude

$$\chi \rightarrow \infty \quad \text{as } T \rightarrow 0!!$$

This is *completely incompatible* with the Third Law.

But Curie's law is a specifically high temperature result (strictly, it applies to the small B/T limit). The general expression for the magnetisation of an ideal paramagnet of N spin $1/2$ moments μ is, Eq. (2.75):

$$M = N\mu \tanh\left(\frac{\mu B}{kT}\right)$$

and corresponding to this, the susceptibility is

$$\chi = \frac{N\mu^2}{VkT} \text{sech}^2\left(\frac{\mu B}{kT}\right).$$

Now we see that

$$\chi \rightarrow 0 \quad \text{as } T \rightarrow 0$$

in conformity with the Third Law, so long as the magnetic field B is finite. Of course, you *have* to use a magnetic field, however small, to measure the susceptibility. Nevertheless, even in the absence of an externally-applied magnetic field, there will be an internal field present: the dipole fields of the magnetic moments themselves. Thus the Third Law is not under threat.

There is a further consideration in the case of fluid systems. In a fluid, where the particles must be treated as delocalised, the statistics will also have an effect. Recall the behaviour of fermions at low temperatures, to be treated in Sec. 2.4. Very roughly, only a fraction T/T_F of the particles are free and available to participate in "normal" behaviour. We then expect that the Curie law behaviour will be modified to

$$\chi \sim \left(\frac{T}{T_F}\right) \times \frac{C}{T}$$

or

$$\chi \sim \frac{C}{T_F}$$

which is indeed a constant, in conformity with the Third Law. This result is correct, but a numerical calculation must be done to determine the numerical constants involved.

1.7.7. *Pessimist's statement of the laws of thermodynamics*

As we have now covered all the laws of thermodynamics we can present their statements in terms of what they prohibit in the operation of Nature.

- **First Law**: You cannot convert heat to work at greater than 100% efficiency
- **Second Law**: You cannot even achieve 100% efficiency — except at $T = 0$.
- **Third Law**: You cannot get to $T = 0$.

This is a simplification, but it encapsulates the underlying truths, and it is easy to remember.

Problems

1.1. Demonstrate that *entropy*, as given by the Boltzmann expression $S = k \ln \Omega$, is an *extensive* property. The best way to do this is to argue *clearly* that Ω is multiplicative.

1.2. Demonstrate that gravitational energy is not extensive: show that the gravitational energy of a sphere of radius r and uniform density varies with volume as V^n and find the exponent n.

1.3. In investigating the conditions for the establishment of equilibrium through the transfer of thermal energy the fundamental requirement is that the entropy of the equilibrium state should be a maximum. Equality of temperature was established from the vanishing of the first derivative of S. What follows from a consideration of the *second derivative*?

1.4. Do particles flow from high μ to low μ or *vice versa*? Explain your reasoning.

1.5. In the derivation of the Boltzmann factor the entropy of the bath was expanded in powers of the energy of the "system of interest". The higher order terms of the expansion are neglected. Discuss the validity of this.

1.6. The Boltzmann factor could have been derived by expanding Ω rather than by expanding S. In that case, however, the expansion cannot be terminated. Why not?

1.7. Show that $\ln N! = \sum_{n=1}^{N} \ln n$. By approximating this sum by an integral obtain *Stirling's approximation*: $\ln N! \approx N \ln N - N$.

1.8. Show that the Gibbs expression for entropy: $S = -k \sum_j P_j \ln P_j$, reduces to the Boltzmann expression $S = k \ln \Omega$ in the case of an isolated system.

1.9. What is the condition that the geometric progression in the derivation of the Bose–Einstein distribution is convergent?

1.10. Show that the trajectory of a 1d harmonic oscillator is an ellipse in phase space. What would the trajectory be if the oscillator were *weakly* damped.

1.11. Why cannot the evolutionary curve in phase space intersect? You need to demonstrate that the evolution from a point is unique.

1.12. Starting from the expression for the Gibbs factor for a many-particle system, write down the grand partition function Ξ and show how it may be expressed as the product of Ξ_k, the grand partition function for the subsystem comprising particles in the kth single-particle state.

1.13. (Difficult) This problem considers the probability distribution for the energy fluctuations in the canonical ensemble. The *moments* of the energy fluctuations are defined by

$$\sigma_n = \frac{1}{Z} \sum_j (E_j - \varepsilon)^n e^{\beta E_j}$$

where $\beta = -1/kT$ and ε is an arbitrary (at this stage) energy.

Show that

$$Z\sigma_n = e^{\beta\varepsilon}\frac{\partial^n}{\partial\beta^n}\{Ze^{-\beta\varepsilon}\}$$

and use this to prove that the fluctuations in an ideal gas obey a *normal distribution* around ε. (You really need to use a computer algebra system to do this problem.)

1.14. For a single-component system with a variable number of particles, the Gibbs free energy is a function of temperature, pressure and number of particles: $G = G(T, p, N)$. Since N is the only extensive variable upon which G depends, show that the chemical potential for this system is equal to the Gibbs free energy per particle: $G = N\mu$.

1.15. Use the definition of the Gibbs free energy together with the result of the previous question to obtain the Euler relation of Appendix 1.

1.16. The energy of a harmonic oscillator may be written as $m\omega^2 x^2/2 + p^2/2m$ so it is quadratic in both position and momentum — thus, classically, equipartition should apply. The energy levels of the quantum harmonic oscillator are given by $\varepsilon_n = \left(\frac{1}{2} + n\right)\hbar\omega$. Show that the partition function of this system is given by

$$Z = \frac{1}{2}\mathrm{cosech}\frac{\hbar\omega}{2kT}$$

and that the internal energy is given by

$$E = \frac{1}{2}\hbar\omega \coth\frac{\hbar\omega}{2kT} = \frac{\hbar\omega}{e^{\hbar\omega/kT} - 1} + \frac{\hbar\omega}{2}.$$

Show that at high temperatures E may be expanded as

$$E = kT + \frac{\hbar^2\omega^2}{12kT} + \cdots$$

Identify the terms in this expansion.

References

[1] R. Bowley and M. Sànchez, *Introductory Statistical Mechanics*, 2nd ed. (Oxford University Press, 1999).

[2] M. W. Zemansky and R. H. Dittman, *Heat and Thermodynamics* (McGraw-Hill, 1968).

[3] P. A. Schilpp (Ed.), *Albert Einstein: Philosopher-Scientist* (Open Court Publishing Co., 1949). In Autobiographical Notes.

[4] T. L. Hill, *Introduction to Statistical Thermodynamics* (Addison Wesley, 1960).

5 L. D. Landau and E. M. Lifshitz, *Statistical Physics* (Pergamon Press, 1980), in the Preface.

6 C. Tsallis, Nonextensive statistics: Theoretical, experimental and computational evidences and connections, *Brazilian J. Phys.* **29** (1999) 1.

7 C. J. Adkins, *Equilibrium Thermodynamics*, 3rd ed. (Cambridge University Press, 1983).

8 A. B. Pippard, *Elements of Classical Thermodynamics* (Cambridge University Press, 1966).

9 H. B. Callen, *Thermodynamics and an Introduction to Thermostatistics*, 2nd ed. (John Wiley, 1985).

10 F. Reif, *Fundamentals of Statistical and Thermal Physics* (McGraw-Hill, 1965).

11 R. P. Feynman, *Statistical Mechanics* (Benjamin, 1972).

12 R. P. Feynman, The reason for antiparticles, in *Elementary Particles and the Laws of Physics — The 1986 Dirac Memorial Lectures*, ed. S. Weinberg (Cambridge University Press, 1987).

13 S. Tomonaga, *The Story of Spin* (University of Chicago Press, 1997).

14 E. Merzbacher, *Quantum Mechanics* (John Wiley, 1970).

15 R. C. Tolman, *The Principles of Statistical Mechanics* (Oxford University Press, 1938).

16 J. S. Dugdale, *Entropy and its Physical Meaning* (Taylor and Francis, 1996).

17 A. J. Leggett, On the minimum entropy of a large system at low temperatures, *Ann. Phys. N.Y.* **72** (1972) 80–106.

18 F. Pobel, *Matter and Methods at Low Temperatures* (Springer-Verlag, 1992).

CHAPTER TWO

PRACTICAL CALCULATIONS WITH IDEAL SYSTEMS

2.1. The Density of States

2.1.1. *Non-interacting systems*

The concept of a non-interacting system is hypothetical since a system of truly non-interacting components could not achieve thermal equilibrium. Thus we are really interested in assemblies of very weakly interacting systems. To be precise, the interactions must be sufficient to lead to thermal equilibrium, but weak enough that these interactions have negligible effect on the energy of the individual particles. We note parenthetically, that the *rate* at which an equilibrium state is established will depend on the strength of the interactions; this is the subject of *non-equilibrium statistical mechanics*. Some aspects will be touched upon in Chapter 5.

Since the single-particle states have well-defined energies, it follows that one can obtain full thermodynamic information about such a system once the energies of the single-particle states and the mean number of particles in each state is known. Then all thermodynamic properties are found by performing sums over states and distribution functions. And for infinite, or very large systems, such sums may usually be converted to integrals.

2.1.2. *Converting sums to integrals*

In quantum statistical mechanics there are many sums over states to be evaluated. An example is the partition function

$$Z(N, V, T) = \sum_i e^{-E_i(N, V)/kT}.$$

It is often convenient to approximate the sums by integrals. And since the individual states are densely packed, negligible error is introduced in so doing. Now if $g(\varepsilon)$ is the number of states with energy between ε and $\varepsilon + d\varepsilon$ then the sum may be approximated by

$$\sum_i e^{-E_i(N,V)/kT} \rightarrow \int_0^\infty g(\varepsilon)e^{-\varepsilon/kT}d\varepsilon.$$

Here $g(\varepsilon)$ is referred to as the (*energy*) *density of states*. If we are studying the properties of a gas then the microstates to be considered are the quantum states of a "particle in a box". And the density of states for a particle in a box may be evaluated in the following way.

2.1.3. *Enumeration of states*

We consider a cubic box of volume V. Each side has length $V^{1/3}$. Elementary quantum mechanics tells us that the wave function of a particle in the box must go to zero at the boundary walls; only standing waves are allowed.

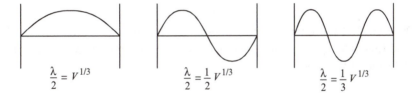

$$\frac{\lambda}{2} = V^{1/3} \qquad \frac{\lambda}{2} = \frac{1}{2}V^{1/3} \qquad \frac{\lambda}{2} = \frac{1}{3}V^{1/3}$$

Fig. 2.1.   Standing waves in a box.

In the general case, the allowed wavelengths λ satisfy $\lambda/2 = V^{1/3}/n$, that is

$$\lambda_n = \frac{2}{n}V^{1/3}, \quad n = 1, 2, 3, 4, \ldots, \infty.$$

In three dimensions there will be a λ for the x, y and z directions:

$$\lambda_{n_x} = 2\frac{V^{1/3}}{n_x}, \quad \lambda_{n_y} = 2\frac{V^{1/3}}{n_y}, \quad \lambda_{n_z} = 2\frac{V^{1/3}}{n_z}.$$

Or, since this corresponds to the components of the wave vector $k_x = 2\pi/\lambda_{n_x}$, etc.,

$$k_x = \frac{\pi}{V^{1/3}}n_x, \quad k_y = \frac{\pi}{V^{1/3}}n_y, \quad k_z = \frac{\pi}{V^{1/3}}n_z. \qquad (2.1)$$

We can now use the deBroglie relation $\mathbf{p} = \hbar\mathbf{k}$ to obtain the momentum and hence the energy.

$$p_x = \frac{\pi\hbar}{V^{1/3}}n_x, \quad p_y = \frac{\pi\hbar}{V^{1/3}}n_y, \quad p_z = \frac{\pi\hbar}{V^{1/3}}n_z.$$

And so for a free particle, the energy is then

$$\varepsilon = \frac{p^2}{2m} = \frac{p_x^2 + p_y^2 + p_z^2}{2m}$$

which is

$$\varepsilon = \frac{\pi^2\hbar^2}{2mV^{2/3}}\left(n_x^2 + n_y^2 + n_z^2\right). \tag{2.2}$$

In this expression it is the triple of quantum numbers (n_x, n_y, n_z) which specify the quantum state. Now each triple defines a point on a cubic grid. If we put

$$R^2 = n_x^2 + n_y^2 + n_z^2$$

then the energy is given by

$$\varepsilon = \frac{\pi^2\hbar^2 R^2}{2mV^{2/3}}.$$

Observe that the energy levels depend on the size of the container. We used this fact in the considerations of Sec. 1.2.2 on the probabilistic interpretation of the First Law.

2.1.4. *Counting states*

Now the number of states of energy up to ε, denoted by $N(\varepsilon)$, is given by the number of points in the octant up to $\varepsilon(R)$. (An octant is used since n_x, n_y and n_z are restricted to being positive.) And the number of points in the octant is approximately equal to the volume of the octant:

$$N(\varepsilon) \frac{1}{8}\frac{4}{3}\pi R^3.$$

But since

$$R = \left(\frac{2mV^{2/3}}{\pi^2\hbar^2}\right)^{1/2}\varepsilon^{1/2},$$

we then obtain

$$N(\varepsilon) = \frac{1}{6}\frac{V}{\pi^2\hbar^3}(2m\varepsilon)^{3/2}.$$

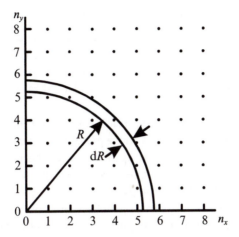

Fig. 2.2.   Counting of quantum states.

Recall that the density of states $g(\varepsilon)$ is defined by saying that the number of states with energy between ε and $\varepsilon + d\varepsilon$ is $g(\varepsilon)$. In other words

$$g(\varepsilon)d\varepsilon = N(\varepsilon + d\varepsilon) - N(\varepsilon)$$

or, simply

$$g(\varepsilon) = \frac{dN(\varepsilon)}{d\varepsilon}.$$

So differentiating $N(\varepsilon)$ we obtain

$$g(\varepsilon) = \frac{1}{4}\frac{V}{\pi^2\hbar^3}(2m)^{3/2}\varepsilon^{1/2} \tag{2.3}$$

which is the required expression for the density of states.

For completeness we shall also give the expressions for the density of states in two and in one dimension. These will be discussed in some of the problems. In a two-dimensional system of area A the density of states is

$$g(\varepsilon) = \frac{mA}{2\pi\hbar^2} \quad \text{(two dimensions);} \tag{2.4}$$

observe this is independent of energy. And in a one-dimensional system of length L the density of states is

$$g(\varepsilon) = \frac{L}{\pi\hbar}\left(\frac{m}{2}\right)^{1/2}\varepsilon^{-1/2} \quad \text{(one dimension).} \tag{2.5}$$

2.1.5.   *General expression for the density of states*

The result of the previous section gives the (energy) density of states for free particles confined to a box of specified volume. This, we argued, was appropriate for the consideration of ideal gases. And it will indeed be quite adequate for the consideration of the classical ideal gas in both the classical and quantum mechanical cases.

In order to be able to treat systems comprising elements other than free particles we need to adopt a more general approach than that of the previous section. Central to that approach was the energy–momentum relation (really the energy-wave vector relation) of the particles: $\varepsilon = p^2/2m = \hbar^2 k^2/2m$. But in the general case this may not be appropriate. Thus at relativistic speeds free particles would obey $\varepsilon = \sqrt{c^2 p^2 + m^2 c^4}$ and extreme-relativistic/massless particles, $\varepsilon = cp$. In these cases as well as those with other energy–momentum relations a more general approach is needed for the density of states.

The key to the general treatment is to consider the density of states in k-space. The point is that the quantisation condition discussed above still applies to the k-states. The wave function for any elements confined to a box must vanish at the walls. This means that the standing wave condition still applies — in other words

$$k_x = \frac{\pi}{V^{1/3}} n_x, \quad k_y = \frac{\pi}{V^{1/3}} n_y, \quad k_z = \frac{\pi}{V^{1/3}} n_z.$$

The allowed states correspond to integer n_x, n_y and n_z, so that there is a uniform density of states in k-space.

The number of states up to a maximum k value is then the volume of the octant

$$N(k) = \frac{1}{8}\frac{4}{3}\pi R^3 = \frac{1}{8}\frac{4}{3}\pi \left(n_x^2 + n_y^2 + n_z^2\right)^{3/2}.$$

And since $n_x = k_x V^{1/3}/\pi$, etc. it follows that

$$N(k) = \frac{V}{6\pi^2} k^3.$$

The density of states in k-space, which we shall denote by $g(k)$, is then the derivative of $N(k)$, so that

$$g(k) = \frac{V}{2\pi^2} k^2.$$

Recall that the energy density of states is defined as

$$g(\varepsilon) = \frac{dN}{d\varepsilon}.$$

And this can be expressed in terms of $g(k)$ by using the chain rule for differentiation

$$g(\varepsilon) = \frac{dN}{d\varepsilon} = \frac{dN}{dk}\frac{dk}{d\varepsilon} = g(k) \left/ \frac{d\varepsilon}{dk} \right. .$$

Thus we have the general result

$$g(\varepsilon) = \frac{V}{2\pi^2} k^2 \left/ \frac{d\varepsilon}{dk} \right. \tag{2.6}$$

where we must eliminate k in favour of ε. Thus to find the energy density of states in this general case we need to know the energy–momentum (energy-wave vector) relation for the particles or excitations.

For completeness we shall also give the general expressions for the density of states in two and in one dimensions. In a two-dimensional system of area A the general density of states is

$$g(\varepsilon) = \frac{A}{2\pi} k \left/ \frac{d\varepsilon}{dk} \right. \quad \text{(two dimensions)}. \tag{2.7}$$

And in a one-dimensional system of length L the general density of states is

$$g(\varepsilon) = \frac{L}{\pi} \left/ \frac{d\varepsilon}{dk} \right. \quad \text{(one dimension)}. \tag{2.8}$$

2.1.6. *General relation between pressure and energy*

It is possible to obtain a very general relation between the internal energy and the pressure of a gas of particles. Central to the argument is the understanding, discussed in Sec. 2.1.3 that the energy levels of a particle in a box depend on the volume of the box, Eq. (2.2).

From the differential expression for the First Law, it follows that

$$p = -\left. \frac{\partial E}{\partial V} \right|_S .$$

We may write the internal energy in terms of the energy levels ε_j and the occupation numbers of these states n_j:

$$E = \sum_j n_j \varepsilon_j .$$

And the differential of this expression is

$$dE = \sum_j n_j d\varepsilon_j + \sum_j \varepsilon_j dn_j .$$

These expressions should be compared with those in the discussion of the probabilistic interpretation of the First Law in Sec. 1.2.2. There, we used the Gibbs approach to the calculation of mean values; here we use the Boltzmann approach. Apart from that they are equivalent.

The second term of the differential expression vanishes at constant entropy. Thus we identify

$$p = -\left.\frac{\partial E}{\partial V}\right|_s = -\sum_j P_j \frac{dE_j}{dV}.$$

Now the energy levels depend on V as a simple power. We shall write

$$\varepsilon_j = AV^n$$

so that

$$\frac{d\varepsilon_j}{dV} = AnV^{n-1}$$

which is convenient to be written as

$$\frac{d\varepsilon_j}{dV} = \frac{AnV^n}{V} = n\frac{\varepsilon_j}{V}.$$

Thus the expression for the pressure becomes

$$p = -\sum_j n_j \frac{d\varepsilon_j}{dV} = -\frac{n}{V}\sum_j n_j \varepsilon_j = -n\frac{E}{V}.$$

For (non-relativistic) particles of mass m (so that the kinetic energy is $p^2/2m$) we saw, in Sec. 2.1.3, Eq. (2.2), that $n = -2/3$. So in this case

$$p = \frac{2}{3}\frac{E}{V} \tag{2.9}$$

or

$$pV = \frac{2}{3}E. \tag{2.10}$$

It is important to appreciate the generality of this result. It relies only on the power law dependence of the energy levels upon volume. It applies to quantum particles irrespective of statistics and thus it applies to classical particles as well. It only holds, however, for non-interacting particles; with interactions the energy levels will depend on volume in a much more complicated way.

As a further example we can consider a gas of ultra-relativistic/massless particles. In this case the energy–momentum relation $\varepsilon = cp$ leads to the energy levels

$$\varepsilon = \frac{c\pi\hbar}{V^{1/3}}\left(n_x^2 + n_y^2 + n_z^2\right)^{1/2} \tag{2.11}$$

so now the index n is $-1/3$ and

$$pV = \frac{1}{3}E. \tag{2.12}$$

This result applies to a gas of photons, and in that case the expression gives the radiation pressure as one third of the energy density. The identical expression for radiation pressure is conventionally derived in electromagnetism[1] — without the need for quantum arguments.

2.2. Identical Particles

2.2.1. *Indistinguishability*

Since the partition function is proportional to probabilities it follows that for composite systems the partition function is a product of the partition functions for the individual subsystems. The free energy is proportional to the logarithm of the partition function and this leads to the extensive variables of composite systems being additive.

In this section we shall examine how the (canonical) partition function of a many-particle system is related to the partition function of a single particle.

If we had an assembly of N identical but *distinguishable* particles the resultant partition function would be the product of the N (same) partition functions of a single particle, z

$$Z = z^N. \tag{2.13}$$

The key question is that of *indistinguishability* of the atoms or molecules of a many-body system. When two identical molecules are interchanged the system is still in the same microstate, so the distinguishable particle result *overcounts* the states in this case. Now the number of ways of redistributing N particles when there are n_1 particles in the first state, n_2 particles in the second state, etc. is

$$\frac{N!}{n_1!\, n_2!\, n_3! \ldots} \tag{2.14}$$

so that for a given distribution $\{n_i\}$ the partition function for identical indistinguishable particles is

$$Z = \frac{n_1!\, n_2!\, n_3! \ldots}{N!} z^N. \tag{2.15}$$

2.2.2. *Classical approximation*

The problem here is the occupation numbers $\{n_i\}$; we do not know these in advance. However at high temperatures the probability of occupancy of any state is small; the probability of multiple occupancy is then negligible. This is the classical régime, where the thermal deBroglie wavelength is very much less than the inter-particle spacing. The thermal deBroglie wavelength is the wavelength, corresponding to the momentum, corresponding to the kinetic energy corresponding to a temperature T. This will appear in Sec. 2.3.1. Under these circumstances the factors $n_1!\ n_2!\ n_3!\ldots$ can be ignored and we have a soluble problem.

In the classical case, we have then

$$Z = \frac{1}{N!} z^N.$$

The Helmholtz free energy

$$F = -kT \ln Z$$

is thus

$$F = -NkT \ln z + kT \ln N!. \tag{2.16}$$

This is N times the Helmholtz free energy for a single particle plus an extra term depending on T and N. So the second term can be ignored so long as we differentiate with respect to something other than T or N. Thus, when differentiating with respect to volume to find the pressure, the result is N times that for a single particle.

The structure of Eq. (2.16) and in particular the logical necessity for the $1/N!$ term in the classical gas partition function will be explored further in Sec. 2.3.4.

2.3. Ideal Classical Gas

2.3.1. *Quantum approach*

The partition function for a single particle is

$$z(V, T) = \sum_i e^{-\varepsilon_i(V)/kT}$$

where ε_i is the energy of the ith single-particle state; these states were enumerated in Sec. 2.1.3. As explained previously, the energy states are closely packed and this allows us to replace the sum over states by an integral using the density of states $g(\varepsilon) = V(2m)^{3/2}\varepsilon^{1/2}/4\pi^2\hbar^3$:

$$z = \int_0^\infty g(\varepsilon)e^{-\varepsilon/kT}d\varepsilon$$

$$= \frac{1}{4}\frac{V}{\pi^2\hbar^3}(2m)^{3/2}\int_0^\infty \varepsilon^{1/2}e^{-\varepsilon/kT}d\varepsilon.$$

Following a change of variable this may be expressed, in terms of a standard integral, as

$$z = \frac{1}{4\pi^2}\left(\frac{2mkT}{\hbar^2}\right)^{3/2}V\int_0^\infty x^{1/2}e^{-x}dx.$$

The integral is $\sqrt{\pi}/2$, so that

$$z = \left(\frac{mkT}{2\pi\hbar^2}\right)^{3/2}V = \frac{V}{\Lambda^3} \tag{2.17}$$

where

$$\Lambda = \sqrt{\frac{2\pi\hbar^2}{mkT}}. \tag{2.18}$$

Now the parameter Λ is known as the thermal deBroglie wavelength; within a numerical factor it is the wavelength corresponding to the momentum corresponding to the thermal kinetic energy of the particle. Thus, in a sense, it represents the quantum "size" of the particle.

For a gas of N particles we then have

$$Z = \frac{1}{N!}z^N = \frac{1}{N!}\left(\frac{V}{\Lambda^3}\right)^N.$$

We use Stirling's approximation, $\ln N! = N\ln N - N$, when evaluating the logarithm:

$$\ln Z = -N\ln N + N + N\ln z = N\ln\left[\left(\frac{mkT}{2\pi\hbar^2}\right)^{3/2}\frac{Ve}{N}\right] \tag{2.19}$$

from which all thermodynamic properties can be found, by the formulae of Sec. 1.4.4.

Since Stirling's approximation becomes true in the thermodynamic limit it follows that by invoking Stirling's approximation we are implicitly taking the thermodynamic limit.

2.3.2. *Classical approach*

It is instructive to consider the calculation of the single-particle partition function from the classical point of view. The classical partition function is given by the integral

$$z = \frac{1}{h^3} \int e^{-\varepsilon/kT} d^3 p d^3 q$$

$$= \frac{1}{8\pi^3 \hbar^3} \int e^{-\varepsilon/kT} d^3 p d^3 q$$

where for the ideal gas $\varepsilon = p^2/2m$. Thus the q integrals are trivial, giving a factor V, and we have

$$z = \frac{V}{8\pi^3 \hbar^3} \left[\int_{-\infty}^{\infty} e^{-p^2/2mkT} dp \right]^3 .$$

The integral is transformed to a pure number by changing variables: $p = x\sqrt{2mkT}$ so that

$$z = \frac{V}{\hbar^3} \left(\frac{mkT}{2\pi^2} \right)^{3/2} \left[\int_{-\infty}^{\infty} e^{-x^2} dx \right]^3 .$$

As in the quantum calculation, the physics is all outside the integral and the integral is just a pure number. The value of the integral is $\sqrt{\pi}$ so that

$$z = \left(\frac{mkT}{2\pi \hbar^2} \right)^{3/2} V$$

just as in the "quantum" calculation; we obtain the identical result. *This justifies the use of h in the normalization factor for the classical state element of phase space.*

2.3.3. *Thermodynamic properties*

In order to investigate the thermodynamic properties of the classical ideal gas we start from the Helmholtz free energy, Eq. (2.19):

$$F = -kT \ln Z = NkT \ln \left[\left(\frac{mkT}{2\pi \hbar^2} \right)^{3/2} \frac{Ve}{N} \right] .$$

Then upon differentiation we obtain

$$p = kT \left. \frac{\partial \ln Z}{\partial V} \right|_{T,N} = NkT \left. \frac{\partial \ln z}{\partial V} \right|_{T} = \frac{NkT}{V} .$$

This is the ideal gas equation, and from this we identify k as Boltzmann's constant. Furthermore, we see that the ideal gas temperature thus corresponds to the "statistical" temperature introduced in the previous chapter.

The internal energy is

$$E = kT^2 \left.\frac{\partial \ln Z}{\partial T}\right|_{V,N} = NkT^2 \frac{d \ln T^{3/2}}{dT} = \frac{3}{2}NkT.$$

This result was obtained previously from equipartition. It also gives another important property of an ideal gas: the internal energy depends *only* on temperature (not pressure or density). This is known as Joule's law. From the energy expression we obtain the thermal capacity

$$C_V = \left.\frac{\partial E}{\partial T}\right|_{V,N} = \frac{3}{2}Nk.$$

It is a constant, independent of temperature, in violation of the Third Law. This is because of the classical approximation — ignoring multiple state occupancy, etc. We also find the entropy and chemical potential:

$$S = Nk \ln \left[\left(\frac{mkT}{2\pi\hbar^2}\right)^{3/2} \frac{V}{N} e^{5/2} \right],$$

$$\mu = -kT \ln \left[\left(\frac{mkT}{2\pi\hbar^2}\right)^{3/2} \frac{V}{N} \right]. \tag{2.20}$$

The formula for the entropy is often expressed as

$$S = Nk \ln V - Nk \ln N + \frac{3}{2} Nk \ln T + Nks_0 \tag{2.21}$$

where the reduced "entropy constant" s_0 is given by

$$s_0 = \frac{3}{2} \ln \left(\frac{mk}{2\pi\hbar^2}\right) + \frac{5}{2}.$$

The above expression for the entropy is known as the Sackur–Tetrode equation. It is often interpreted as indicating different contributions to the entropy: the volume contribution in the first term, the number contribution in the second term and the temperature contribution in the third term. Such an identification is entirely incorrect; this matter is explored in Problem 2.3.

2.3.4.   *The 1/N! term in the partition function*

In Sec. 2.2.2 we argued that the factor $1/N!$ was needed in the partition function of a classical gas of identical particles. In this section, we explore this idea further and we shall see that the factor is required to ensure that the mathematical expression for the Helmholtz free energy is extensive — as it obviously must be.

The partition function for a gas of N classical particles may be written as

$$Z = \frac{1}{N!} z^N$$

$$= \frac{1}{N!} \left(\frac{mkT}{2\pi \hbar^2} \right)^{3N/2} V^N. \tag{2.22}$$

If we use Stirling's approximation for the factorial in the form

$$N! \sim \left(\frac{N}{e} \right)^N$$

(equivalent to the usual expression $\ln N! \sim N \ln N - N$) then the expression for Z becomes

$$Z = \left(\frac{mkT}{2\pi \hbar^2} \right)^{3N/2} e^N \left(\frac{V}{N} \right)^N. \tag{2.23}$$

The Helmholtz free energy is given by

$$F = -kT \ln Z$$

$$= -NkT \ln \left\{ \left(\frac{mkT}{2\pi \hbar^2} \right)^{3/2} e \frac{V}{N} \right\}.$$

Here the argument of the logarithm is intensive because the extensive V is divided by the extensive N. Then the N outside the logarithm ensures that the free energy is an extensive quantity as required.

It is to be expected that the full extensivity of F will only become manifest in the thermodynamic limit — i.e. upon using Stirling's approximation.

The important point to note is that the N inside the logarithm is a consequence of the $1/N!$ in the partition function. Without this factor the N would not be there; then F would not be extensive. This is a further justification of the $1/N!$ factor.

2.3.5. *Entropy of mixing*

Let us consider the mixing of two gases. Figure 2.3 shows a container with a wall separating it into two regions. In the left-hand region there are N_1 particles and in the right there are N_2 particles. We shall assume that both regions are at the same temperature and pressure. Furthermore we will assume that the number density N/V is the same on both sides.

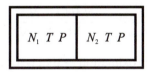

Fig. 2.3. Mixing of gases.

We shall investigate what happens when the wall is removed. In particular, we shall consider what happens when the gases on either side are different and when they are the same.

When the partition is in place the partition function is simply the product of the partition functions of two systems. We use the partition function expression of Eq. (2.3) so that

$$Z = Z_1 Z_2$$
$$= \frac{1}{N_1!} \left(\frac{mkT}{2\pi\hbar^2} \right)^{3N_1/2} V_1^{N_1} \frac{1}{N_2!} \left(\frac{mkT}{2\pi\hbar^2} \right)^{3N_2/2} V_2^{N_2}.$$

This may be simplified to

$$Z = \frac{1}{N_1! N_2!} \left(\frac{mkT}{2\pi\hbar^2} \right)^{3(N_1+N_2)/2} V_1^{N_1} V_2^{N_2}.$$

This is the same whether the particles are identical or different.

Now we shall remove the separating wall and allow the gases to mix. There are two cases to consider: identical and different particles.

(a) Identical particles
If the particles are identical then the new partition function is that corresponding to $N = N_1 + N_2$, $V = V_1 + V_2$. Thus the "after" partition function will be

$$Z = \frac{1}{(N_1+N_2)!} \left(\frac{mkT}{2\pi\hbar^2} \right)^{3(N_1+N_2)/2} (V_1 + V_2)^{(N_1+N_2)}.$$

The ratio of the after to before partition function is

$$\frac{Z_{\text{after}}}{Z_{\text{before}}} = \frac{N_1! N_2!}{(N_1 + N_2)!} \frac{(V_1 + V_2)^{(N_1 + N_2)}}{V_1^{N_1} V_2^{N_2}}.$$

This may be simplified. Since $V_1/N_1 = V_2/N_2 = V/N$, it follows that $V_1 = V N_1/N$ and $V_2 = V N_2/N$ so that

$$\frac{Z_{\text{after}}}{Z_{\text{before}}} = \frac{N_1! N_2!}{N!} \frac{N^N}{N_1^{N_1} N_2^{N_2}}.$$

Upon using Stirling's approximation — that is, upon taking the thermodynamic limit — this expression becomes unity. That is, in the case of identical particles

$$Z_{\text{after}} = Z_{\text{before}}.$$

So for identical particles there is no effect from removing the separating wall and the original (macro-)state will be recovered if the wall is replaced. This is as expected on physical grounds, but the result would not have been obtained if the $1/N!$ term in the partition function had not been used.

(b) Different particles
If the particles on the two sides were different, then on removing the wall we have a composite system with N_1 type-1 particles in the volume $V = V_1 + V_2$ and N_2 type-2 particles in the same volume. Thus we have

$$Z = \frac{1}{N_1!} \left(\frac{mkT}{2\pi\hbar^2} \right)^{3N_1/2} V^{N_1} \frac{1}{N_2!} \left(\frac{mkT}{2\pi\hbar^2} \right)^{3N_2/2} V^{N_2}$$

$$= \frac{1}{N_1! N_2!} \left(\frac{mkT}{2\pi\hbar^2} \right)^{3N/2} V^N.$$

The ratio of the after to the before partition function is then

$$\frac{Z_{\text{after}}}{Z_{\text{before}}} = \frac{V^N}{V_1^{N_1} V_2^{N_2}}.$$

In this case, the "after" partition function is different to the "before". Then, for unlike particles, the free energy and some of the other thermodynamic parameters will change when the wall is removed.

The change in the free energy is given by

$$F_{\text{after}} - F_{\text{before}} = -kT \ln \frac{Z_{\text{after}}}{Z_{\text{before}}}$$

$$= -kT \ln \left\{ \frac{V^N}{V_1^{N_1} V_2^{N_2}} \right\}.$$

This will lead to a change of entropy. Since

$$S = -\left. \frac{\partial F}{\partial T} \right|_{V,N}$$

it follows that the entropy change upon removal of the wall is

$$\Delta S = k \ln \left\{ \frac{V^N}{V_1^{N_1} V_2^{N_2}} \right\}$$

$$= Nk \ln V - N_1 k \ln V_1 - N_2 k \ln V_2 .$$

This may be written as

$$\Delta S = N_1 k \ln \frac{V}{V_1} + N_2 k \ln \frac{V}{V_2} \tag{2.24}$$

which shows that the entropy increases when the wall is removed. This increase of entropy when two different gases are mixed is known as the *entropy of mixing*. The expression may be understood from the Boltzmann entropy expression, using simple counting arguments: the first term gives the entropy increase of the N_1 type-1 particles expanding from volume V_1 into volume V while the second term gives the entropy increase of the N_2 type-2 particles expanding from volume V_2 into volume V.

2.4. Ideal Fermi Gas

2.4.0. *Methodology for quantum gases*

The Bose–Einstein and the Fermi–Dirac distribution functions give the mean number of particles in a given single particle quantum state in terms of the temperature T and the chemical potential μ. The temperature and chemical potential are the *intensive* variables that determine the equilibrium distribution $\bar{n}(\varepsilon)$. Now we have a good intuitive feel for the temperature of a system. But the chemical potential is different. This determines the number of particles in the system when it is in thermal equilibrium with a particle reservoir. In reality, however, it may be more intuitive to speak of a system containing a given (mean) number of particles. In that case, it is the number of particles in the system that determines the chemical potential.

Now the number of particles in the system is given by

$$N = \sum_i \bar{n}(\varepsilon_i)$$

which converts to the integral

$$N = \int_0^\infty \alpha \bar{n}(\varepsilon) g(\varepsilon) d\varepsilon$$

where $g(\varepsilon)$ is the energy density of states, α is the factor which accounts for the degeneracy of the particles' spin states. This is 2 for electrons since there are two spin states for a spin $S = 1/2$; more generally it will be $2S + 1$, (but see Sec. 2.6.3 dealing with photons — massless particles are a special case).

The expression for N is inverted to give μ, which can then be used in the distribution function to find the other properties of the system. For instance, the internal energy of the system would be found from

$$E = \sum_i \varepsilon_i \bar{n}(\varepsilon_i)$$

or, in integral form

$$E = \int_0^\infty \alpha \varepsilon \bar{n}(\varepsilon) g(\varepsilon) d\varepsilon.$$

Thus the methodology for treating a quantum gas starts with the expression for the number of particles N. The expression is inverted to find the chemical potential μ. This is substituted into the distribution function $\bar{n}(\varepsilon)$, from which all other properties may be found.

2.4.1. *Fermi gas at zero temperature*

At zero temperature the Fermi–Dirac distribution function

$$n(\varepsilon) = \frac{1}{e^{(\varepsilon - \mu)/kT} + 1}$$

(for notational convenience we have dropped the bar over n) becomes a box function

$$n(\varepsilon) = 1 \quad \varepsilon < \mu$$
$$= 0 \quad \varepsilon > \mu.$$

Note that in general the chemical potential depends on temperature. Its zero temperature value is called the Fermi energy

$$\varepsilon_F = \mu(T = 0).$$

And the temperature defined by $kT_F = \varepsilon_F$ is known as the Fermi temperature.

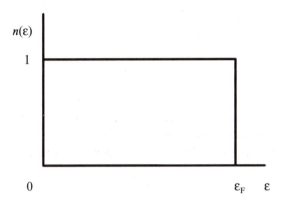

Fig. 2.4. Fermi–Dirac distribution at $T = 0$.

In accordance with the methodology described in the previous section, the first thing to do is to evaluate the total number of particles in the system in order to find the chemical potential — the Fermi energy in the $T = 0$ case.

The density of states is given by Eq. (2.3):

$$g(\varepsilon) = \frac{1}{4} \frac{V}{\pi^2 \hbar^3} (2m)^{3/2} \varepsilon^{1/2}.$$

Where appropriate we shall consider spin $1/2$ particles such as electrons or ^{3}He atoms, when we will take $\alpha = 2$. The total number of particles in the system is

$$N = \frac{\alpha V}{4\pi^2 \hbar^3} (2m)^{3/2} \int_0^{\varepsilon_F} \varepsilon^{1/2} \, d\varepsilon$$

$$= \frac{\alpha V}{6\pi^2 \hbar^3} (2m\varepsilon_F)^{3/2}.$$

And this may be inverted to obtain the Fermi energy

$$\varepsilon_F = \frac{\hbar^2}{2m} \left(\frac{6\pi^2}{\alpha} \frac{N}{V} \right)^{2/3}. \tag{2.25}$$

This gives the chemical potential at zero temperature. Observe that it depends on the *density* of particles in the system N/V, so the Fermi energy is, as expected, an intensive variable.

The Fermi energy provides a useful energy scale for the consideration of the properties of Fermi systems. For this reason, it proves useful and instructive to work in terms of this parameter and we therefore express the density of states in terms of the Fermi energy

$$g(\varepsilon) = \frac{3N}{2\alpha\varepsilon_{\mathrm{F}}^{3/2}} \varepsilon^{1/2}$$

$$= \frac{3}{2\alpha} \frac{N}{\varepsilon_{\mathrm{F}}} \left(\frac{\varepsilon}{\varepsilon_{\mathrm{F}}}\right)^{1/2}. \tag{2.26}$$

Having obtained the zero-temperature chemical potential, the Fermi–Dirac function is now completely specified at $T = 0$, and we can proceed to find the zero temperature internal energy of the system. This is given by

$$E(T = 0) = \frac{3N}{2\varepsilon_{\mathrm{F}}^{3/2}} \int_0^{\varepsilon_{\mathrm{F}}} \varepsilon^{3/2} \, \mathrm{d}\varepsilon,$$

which evaluates to

$$E = \frac{3}{5} N \varepsilon_{\mathrm{F}}. \tag{2.27}$$

The internal energy is proportional to the number of particles in the system and so it is, as expected, an extensive quantity.

This considerable ground state energy is a consequence of the Pauli exclusion principle.

2.4.2. *Fermi gas at low temperatures — simple model*

The effect of a small finite temperature may be modelled very simply by approximating the Fermi–Dirac distribution function by a piecewise-linear function. This must match the slope of the curve at $\varepsilon = \mu$, and the derivative is found to be

$$\left.\frac{\mathrm{d}n(\varepsilon)}{\mathrm{d}\varepsilon}\right|_{\varepsilon=\mu} = \frac{-1}{4kT}. \tag{2.28}$$

This indicates that the energy width of the transition region is $\Delta\varepsilon \sim kT$.

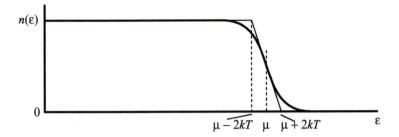

Fig. 2.5.   Simple modelling of Fermi–Dirac distribution.

The distribution function, in this approximation, is then

$$n(\varepsilon) = 1, \qquad\qquad\qquad\qquad 0 < \varepsilon < \mu - 2kT$$
$$= 1 - \frac{\varepsilon - \mu + 2kT}{4kT}, \qquad \mu - 2kT < \varepsilon < \mu + 2kT$$
$$= 0, \qquad\qquad\qquad\qquad \mu + 2kT < \varepsilon$$

from which the thermodynamic properties may be calculated.

The chemical potential may now be found by considering the total number of particles in the system:

$$N = \alpha \int_0^\infty n(\varepsilon) g(\varepsilon) \mathrm{d}\varepsilon$$
$$= \frac{3N}{2\varepsilon_{\mathrm{F}}^{3/2}} \int_0^\infty n(\varepsilon) \varepsilon^{1/2} \, \mathrm{d}\varepsilon.$$

Observe that here the number N appears on both sides and so it cancels. The reason for this is that we have chosen to express the density of states in terms of the Fermi energy. Thus, while the N cancels from both sides the number (density) is contained in the Fermi energy. So in this approach we are really expressing the Fermi energy in terms of the chemical potential:

$$\varepsilon_{\mathrm{F}}^{3/2} = \frac{3}{2} \int_0^\infty n(\varepsilon) \varepsilon^{1/2} \, \mathrm{d}\varepsilon$$

and we wish to invert this relation to give the chemical potential in terms of the Fermi energy (and temperature).

Using the approximate form for $n(\varepsilon)$, upon integration, this gives

$$\varepsilon_F^{3/2} = \frac{3}{2}\left(\frac{2}{3}\mu^{3/2} + \frac{1}{3}k^2T^2\mu^{-1/2} + \frac{1}{20}k^4T^4\mu^{-5/2} + \cdots\right),$$

or, on simplifying:

$$\varepsilon_F^{3/2} = \mu^{3/2}\left\{1 + \frac{1}{2}\left(\frac{kT}{\mu}\right)^2 + \frac{3}{40}\left(\frac{kT}{\mu}\right)^4 + \cdots\right\}.$$

The Fermi energy is then the 2/3 power of this

$$\varepsilon_F = \mu\left\{1 + \frac{1}{2}\left(\frac{kT}{\mu}\right)^2 + \frac{3}{40}\left(\frac{kT}{\mu}\right)^4 + \cdots\right\}^{2/3},$$

which, by the binomial theorem, gives ε_F as

$$\varepsilon_F = \mu\left\{1 + \frac{1}{3}\left(\frac{kT}{\mu}\right)^2 + \frac{1}{45}\left(\frac{kT}{\mu}\right)^4 + \cdots\right\}$$

$$= \mu + \frac{1}{3}\frac{(kT)^2}{\mu} + \frac{1}{45}\frac{(kT)^4}{\mu^3} + \cdots.$$

This may be inverted to give μ in terms of ε_F:

$$\mu = \varepsilon_F\left\{1 - \frac{1}{3}\left(\frac{kT}{\varepsilon_F}\right)^2 - \frac{2}{15}\left(\frac{kT}{\varepsilon_F}\right)^4 + \cdots\right\}. \qquad (2.29)$$

This shows that as the temperature is increased from $T= 0$ the chemical potential decreases from its zero temperature value, and the first term is in T^2.

In a similar way the internal energy is found to be

$$E = E_0\left\{1 + \frac{5}{3}\left(\frac{kT}{\varepsilon_F}\right)^2 - \frac{2}{3}\left(\frac{kT}{\varepsilon_F}\right)^4 + \cdots\right\} \qquad (2.30)$$

up to the term in T^4.

We should note that the approximation of the Fermi–Dirac distribution used in this section provides only a model that allows the simple calculation of properties of the Fermi gas, indicating their general behaviour at low temperatures. But the coefficients of the powers of kT/ε_F are incorrect; this is simply a model that permits demonstration of the *qualitative* low temperature behaviour.

2.4.3. *Fermi gas at low temperatures — series expansion*

We do have formal expressions for the exact thermodynamic behaviour of a gas of fermions. In the previous section, a model was described which treated the deviations from $T = 0$ behaviour in a simplistic and qualitative way. What is required is a systematic procedure for calculating finite (but low) temperature behaviour. We shall see that such a procedure is possible based on the special shape of the Fermi–Dirac distribution function; this approach was pioneered by Arnold Sommerfeld.[2]

Very generally, one requires to evaluate integrals of the form

$$I = \int_0^\infty n(\varepsilon)\varphi(\varepsilon)d\varepsilon$$

where $n(\varepsilon)$ is the Fermi–Dirac distribution function and $\varphi(\varepsilon)$ is the function to be integrated over, including the density of states. If we define $\psi(\varepsilon)$, the integral of $\varphi(\varepsilon)$, by

$$\psi(\varepsilon) = \int \varphi(\varepsilon)d\varepsilon$$

then the expression for I may be integrated by parts to give

$$I = n(\varepsilon)\psi(\varepsilon)\big|_0^\infty - \int_0^\infty n'(\varepsilon)\psi(\varepsilon)d\varepsilon.$$

Now since $n(\infty) = 0$ from the form of the Fermi–Dirac function, and $\psi(0) = 0$ from its definition, the first term vanishes and one is left with

$$I = -\int_0^\infty n'(\varepsilon)\psi(\varepsilon)d\varepsilon. \tag{2.31}$$

In evaluating this integral, the important point to note is that $n'(\varepsilon)$ is sharply peaked at $\varepsilon = \mu$, particularly at the lowest of temperatures; this was Sommerfeld's key point.

It then follows that an efficiently convergent series for I may be obtained by expanding $\psi(\varepsilon)$ about $\varepsilon = \mu$ and integrating Eq. (2.31) term by term. The expansion is

$$\psi(\varepsilon) = \sum_{n=0}^\infty \frac{1}{n!} \frac{d^n\psi}{d\varepsilon^n}\bigg|_{\varepsilon=\mu} (\varepsilon - \mu)^n$$

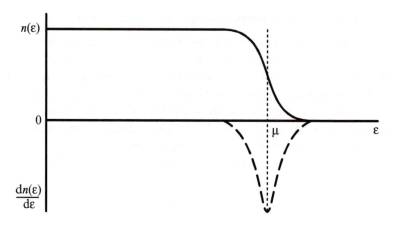

Fig. 2.6.   Fermi–Dirac distribution and its derivative.

so that

$$I = -\sum_{n=0}^{\infty} \frac{1}{n!} \frac{d^n \psi}{d\varepsilon^n}\bigg|_{\varepsilon=\mu} \int_0^{\infty} n'(\varepsilon)(\varepsilon - \mu)^n \, d\varepsilon.$$

The integral is evaluated through the substitution $x = (\varepsilon - \mu)/kT$, and since n' goes to zero away from $\varepsilon = \mu$ the lower limit of the integral can be extended to $-\infty$ without introducing any significant error. Then

$$\int_0^{\infty} n'(\varepsilon)(\varepsilon - \mu)^n d\varepsilon = -(kT)^n I_n$$

where

$$I_n = \int_{-\infty}^{\infty} \frac{e^x}{(e^x + 1)^2} x^n dx. \tag{2.32}$$

Since $e^x/(e^x + 1)^2 = 2/\cosh(x/2)$ is an even function, I_n for odd n vanish. We then find

$$I = \sum_{n=0}^{\infty} I_n \frac{(kT)^n}{n!} \frac{d^n \psi}{d\varepsilon^n}\bigg|_{\varepsilon=\mu}$$

which is a series expansion for the quantity I in powers of temperature. This is thus a *low temperature* expansion. By contrast, the high temperature behaviour will be treated in Sec. 2.4.5.

The general expression for the integrals is

$$I_n = (2 - 2^{2-n})\zeta(n)n!$$

or equivalently

$$I_n = (-1)^{n/2}(2 - 2^n)\pi^n B_n$$

for even n; here $\zeta(n)$ is Riemann's zeta function and B_n is the nth Bernoulli number. Thus we may write I as

$$I = \sum_{\substack{n=0 \\ \text{even } n}}^{\infty} (2 - 2^{2-n})\zeta(n)\,(kT)^n \left.\frac{d^n\psi}{d\varepsilon^n}\right|_{\varepsilon=\mu}.$$

The first few I_n values are

$$I_0 = 1$$
$$I_2 = \pi^2/3$$
$$I_4 = 7\pi^4/15$$
$$I_6 = 31\pi^6/21$$
$$I_8 = 127\pi^8/15$$
$$I_{10} = 2555\pi^{10}/33,$$

so that

$$I = \psi(\mu) + \frac{\pi^2}{6}(kT)^2 \left.\frac{d^2\psi}{d\varepsilon^2}\right|_{\varepsilon=\mu} + \frac{7\pi^4}{360}(kT)^4 \left.\frac{d^4\psi}{d\varepsilon^4}\right|_{\varepsilon=\mu} + \cdots$$

or, in terms of the original function $\varphi(\varepsilon)$,

$$I = \int_0^\mu \varphi(\varepsilon)d\varepsilon + \frac{\pi^2}{6}(kT)^2 \left.\frac{d\varphi}{d\varepsilon}\right|_{\varepsilon=\mu} + \frac{7\pi^4}{360}(kT)^4 \left.\frac{d^3\varphi}{d\varepsilon^3}\right|_{\varepsilon=\mu} + \cdots. \qquad (2.33)$$

Note this is not quite the simple power series it appears since the chemical potential μ also depends upon temperature. Thus chemical potential must always be found before proceeding to the other thermodynamic quantities.

We should also point out that in extending the integrals in Eq. (2.32) to minus infinity, we are neglecting exponentially small terms. Landau and Lifshitz[3] pointed out that the resultant expansion is thus asymptotic and not a convergent series. This is dramatically illustrated in the two-dimensional case, considered in Problem 2.6.

These series expansions are conveniently evaluated using a computer symbolic mathematics system such as *Mathematica*. It is then expedient to recast the expression for I as

$$I = -\int_0^\infty n'(\varepsilon)\psi(\varepsilon)\mathrm{d}\varepsilon$$

$$= \frac{1}{kT}\int_0^\infty \frac{e^{(\varepsilon-\mu)/kT}}{\left(e^{(\varepsilon-\mu)/kT}+1\right)^2}\psi(\varepsilon)\mathrm{d}\varepsilon.$$

Then making the substitution $x = (\varepsilon - \mu)/kT$, and extending the lower limit of the integral to $-\infty$ as explained above,

$$I = \int_{-\infty}^\infty \frac{e^x}{(e^x+1)^2}\psi(kTx+\mu)\mathrm{d}x.$$

The procedure for evaluating the low temperature series for I then comprises expanding ψ about μ in powers of x and integrating term by term.

Chemical potential

The chemical potential may now be found by considering the total number of particles in the system:

$$N = \alpha \int_0^\infty n(\varepsilon)g(\varepsilon)\mathrm{d}\varepsilon$$

$$= \frac{3N}{2\varepsilon_F^{3/2}}\int_0^\infty n(\varepsilon)\varepsilon^{1/2}\,\mathrm{d}\varepsilon.$$

Observe that here the number N appears on both sides and so it cancels. The reason for this is that we have chosen to express the density of states in terms of the Fermi energy. Thus while N cancels from both sides the number (density) is contained in the Fermi energy. So in this approach we are really expressing the Fermi energy in terms of the chemical potential

$$\varepsilon_F^{3/2} = \frac{3}{2}\int_0^\infty n(\varepsilon)\varepsilon^{1/2}\,\mathrm{d}\varepsilon \tag{2.34}$$

and we wish to invert this relation to give the chemical potential in terms of the Fermi energy (and temperature).

In this case

$$\varphi(\varepsilon) = \frac{3N}{2\varepsilon_F^{3/2}} \varepsilon^{1/2}$$

and

$$\psi(\varepsilon) = \frac{N}{\varepsilon_F^{3/2}} \varepsilon^{3/2}.$$

Then using the above result in Eq. (2.8) or using the *Mathematica* procedure in the Appendix, we find

$$\varepsilon_F^{3/2} = \mu^{3/2} \left\{ 1 + \frac{\pi^2}{6} \frac{3}{4} \left(\frac{kT}{\mu}\right)^2 + \frac{7\pi^4}{360} \frac{9}{16} \left(\frac{kT}{\mu}\right)^4 + \cdots \right\},$$

or, upon simplification:

$$\varepsilon_F^{3/2} = \mu^{3/2} \left\{ 1 + \frac{1}{8}\pi^2 \left(\frac{kT}{\mu}\right)^2 + \frac{7}{640}\pi^4 \left(\frac{kT}{\mu}\right)^4 + \cdots \right\}.$$

The Fermi energy is then the $2/3$ power of this

$$\varepsilon_F = \mu \left\{ 1 + \frac{1}{8}\pi^2 \left(\frac{kT}{\mu}\right)^2 + \frac{7}{640}\pi^4 \left(\frac{kT}{\mu}\right)^4 + \cdots \right\}^{2/3}$$

which, by the binomial theorem, gives ε_F as

$$\varepsilon_F = \mu \left\{ 1 + \frac{2}{3}\frac{1}{8}\pi^2 \left(\frac{kT}{\mu}\right)^2 + \frac{1}{180}\pi^4 \left(\frac{kT}{\mu}\right)^4 + \cdots \right\}$$

$$= \mu + \frac{\pi^2}{12}\frac{(kT)^2}{\mu} + \frac{\pi^4}{180}\frac{(kT)^4}{\mu^3} + \cdots$$

and to the same order this may be inverted to give

$$\mu = \varepsilon_F \left\{ 1 - \frac{\pi^2}{12} \left(\frac{kT}{\varepsilon_F}\right)^2 - \frac{\pi^4}{80} \left(\frac{kT}{\varepsilon_F}\right)^4 + \cdots \right\}. \qquad (2.35)$$

This shows that as the temperature is increased from $T = 0$ the chemical potential decreases from its zero temperature value, and the first term is in T^2. The procedure can be extended easily to find higher powers of T in the expansion; indeed the *Mathematica* code in Appendix 3 can do this automatically.

Internal energy

The internal energy for the gas of fermions is given by:

$$E = \alpha \int_0^\infty \varepsilon n(\varepsilon) g(\varepsilon) d\varepsilon$$

$$= \frac{3N}{2\varepsilon_F^{3/2}} \int_0^\infty n(\varepsilon) \varepsilon^{3/2} d\varepsilon,$$

so that in this case

$$\varphi(\varepsilon) = \frac{3N}{2\varepsilon_F^{3/2}} \varepsilon^{3/2}$$

and

$$\psi(\varepsilon) = \frac{3}{5} \frac{N}{\varepsilon_F^{3/2}} \varepsilon^{5/2}.$$

Then using the above result in Eq. (2.33) or using the *Mathematica* procedure in the Appendix, we find

$$E = \frac{3}{5} \frac{N\mu^{5/2}}{\varepsilon_F^{3/2}} + \frac{\pi^2}{6}(kT)^2 \frac{9}{4} \frac{N\mu^{1/2}}{\varepsilon_F^{3/2}} - \frac{7\pi^4}{360}(kT)^4 \frac{9}{16} \frac{N\mu^{-3/2}}{\varepsilon_F^{3/2}} + \cdots$$

This, however, gives the energy in terms of μ, which has a temperature dependence of its own. Thus we must substitute for μ, from Eq. (2.35). This will then give the temperature dependence of the internal energy. Up to the term in T^4 it is:

$$E = \frac{3}{5} N\varepsilon_F + \frac{\pi^2}{4} N\varepsilon_F \left(\frac{kT}{\varepsilon_F}\right)^2 - \frac{3\pi^4}{80} N\varepsilon_F \left(\frac{kT}{\varepsilon_F}\right)^4 + \cdots$$

$$= E_0 \left\{ 1 + \frac{5\pi^2}{12} \left(\frac{kT}{\varepsilon_F}\right)^2 - \frac{5\pi^4}{80} \left(\frac{kT}{\varepsilon_F}\right)^4 + \cdots \right\}. \tag{2.36}$$

This procedure can be extended easily to find higher powers of T in the expansion; indeed the *Mathematica* code in the Appendix can do this automatically.

Thermal capacity

The thermal capacity is found by differentiating the internal energy with respect to temperature:

$$C_V = \left.\frac{\partial E}{\partial T}\right|_V$$

so that

$$C_V = Nk\left\{\frac{\pi^2}{2}\frac{kT}{\varepsilon_F} - \frac{3\pi^4}{20}\left(\frac{kT}{\varepsilon_F}\right)^3 + \cdots\right\}. \tag{2.37}$$

We see that the heat capacity goes to zero with temperature, in agreement with the Third Law. And in this case it goes to zero linearly. In this limit we observe that C_V is $\frac{\pi^2}{3}\frac{kT}{\varepsilon_F}$ times the classical value.

2.4.4. *More general treatment of low temperature heat capacity*

The low temperature heat capacity of real fermions, such as electrons or liquid ^{3}He, is observed to follow the linear temperature behaviour above, but the coefficient can be different from that calculated from Eq. (2.37)

$$C_V = Nk^2T\frac{\pi^2\hbar^2}{4m}\left(\frac{\alpha V}{6\pi^2 N}\right)^{2/3},$$

where we have substituted for the Fermi energy. The simplest interpretation of this observation is to ascribe to the system an effective mass, in terms of which the above equation becomes rehabilitated.

At a deeper level the explanation is that for such systems the dispersion relation connecting the energy and the momentum is not that appropriate for free particles and so the density of states may be different from the form used above. For this reason it is instructive to consider the heat capacity of a system of fermions where the form for the density of states function is not assumed. This means we are no longer considering non-interacting fermions and so strictly this is a topic for the next chapter. However we shall place the discussion here as it follows naturally from our previous considerations. In order to proceed we shall utilise a generalisation of the Sommerfeld expansion method. Our treatment was inspired by the discussion of Reif.[4]

In accordance with our general procedures, the first thing to do is to obtain a suitable expression for the chemical potential for the system. Since we are considering low temperatures, we know that μ will be close to the Fermi energy, but we will require the small deviation from this value.

The number of particles in the system is given by

$$N = \alpha \int_0^\infty n(\varepsilon)g(\varepsilon)d\varepsilon$$

where here $g(\varepsilon)$ is our general expression for the density of states. In terms of the Sommerfeld expansion, Eq. (2.8), here

$$\varphi(\varepsilon) = \alpha g(\varepsilon)$$

so that up to second-order

$$N = \alpha \int_0^\mu g(\varepsilon)d\varepsilon + \alpha \frac{\pi^2}{6}(kT)^2 g'(\varepsilon).$$

In this expression we shall split the integral in the first term into the range from 0 to ε_F and the range from ε_F to μ:

$$N = \alpha \int_0^{\varepsilon_F} g(\varepsilon)d\varepsilon + \alpha \int_{\varepsilon_F}^\mu g(\varepsilon)d\varepsilon + \alpha \frac{\pi^2}{6}(kT)^2 g'(\varepsilon).$$

Here the first integral is simply the total number of particles N. In the second integral we note that ε_F is close to μ and that over this small range the argument of the integral, $g(\varepsilon)$ will hardly change. Thus the second term may be approximated by

$$\alpha(\mu - \varepsilon_F)g(\varepsilon_F).$$

Since the third term already has a T^2 factor, we introduce negligible error here by replacing ε by ε_F. Then we have

$$(\mu - \varepsilon_F)g(\varepsilon_F) + \frac{\pi^2}{6}(kT)^2 g'(\varepsilon) = 0$$

so that to this order of approximation

$$\mu - \varepsilon_F = -\frac{\pi^2}{6}(kT)^2 \frac{g'(\varepsilon)}{g(\varepsilon_F)}. \qquad (2.38)$$

This gives the leading order deviation of the chemical potential from the zero-temperature value of the Fermi energy.

Now moving to the calculation of the internal energy, in this case we have

$$\varphi(\varepsilon) = \alpha \varepsilon g(\varepsilon).$$

Then the series for the energy may be written in a similar way as

$$E = \alpha \int_0^{\varepsilon_F} \varepsilon g(\varepsilon) d\varepsilon + \alpha \int_{\varepsilon_F}^{\mu} \varepsilon g(\varepsilon) d\varepsilon + \alpha \frac{\pi^2}{6} (kT)^2 \{ g(\varepsilon) + \varepsilon g'(\varepsilon) \}.$$

Here the first term is the ground state energy E_0. In the second term ε_F is again close to μ and over this small range the argument of the integral, $\varepsilon g(\varepsilon)$ will hardly change. Thus the second term may be approximated by

$$\alpha(\mu - \varepsilon_F)\varepsilon_F g(\varepsilon_F).$$

But now we have an expression for the $\mu - \varepsilon_F$ in Eq. (2.38). Thus the second term may be written as

$$-\alpha \frac{\pi^2}{6} (kT)^2 g'(\varepsilon_F)\varepsilon_F.$$

And since the third term already has a T^2 factor we introduce negligible error by replacing ε by ε_F. Then we have

$$E = E_0 - \alpha \frac{\pi^2}{6} (kT)^2 g'(\varepsilon_F)\varepsilon_F + \alpha \frac{\pi^2}{6} (kT)^2 \{ g(\varepsilon_F) + \varepsilon_F g'(\varepsilon_F) \}$$

which simplifies to

$$E = E_0 + \alpha \frac{\pi^2}{6} (kT)^2 g(\varepsilon_F)$$

as the derivative terms cancel. This result shows that the internal energy depends on the density of states only at the Fermi surface.

The heat capacity is found by differentiating the internal energy. So in the low temperature limit the heat capacity of an assembly of fermions with arbitrary dispersion relation and thus arbitrary density of states is

$$C_V = \frac{1}{3}\alpha \pi^2 g(\varepsilon_F) k^2 T. \tag{2.39}$$

We see that the low temperature heat capacity remains linear in temperature, and we see that it is determined by the density of states at the Fermi surface.

2.4.5.   *High temperature behaviour — the classical limit*

We now turn our attention to the behaviour of a Fermi gas at high temperatures. When the temperature is high, the probability of occupation of the individual single-particle states becomes significantly less than unity. Then the Fermi distribution

$$n(\varepsilon) = \frac{1}{e^{(\varepsilon-\mu)/kT} + 1}$$

is much less than unity; the denominator is much greater than unity; the exponential is then large compared with $+1$ in the denominator. So in that limit that $+1$ in the denominator becomes negligible; the distinction between fermions and bosons — and indeed classical particles (maxwellons) — disappears. You should recall our discussion in Sec. 2.2.2, where we argued that quantum effects become insignificant when the probability of occupation of the states is small. *Mathematica* notebooks for these calculations are given in Appendixes 3.3 and 3.4.

In this limit we shall treat the Bose, Fermi and classical cases together, by writing the distribution function as

$$n(\varepsilon) = \frac{1}{e^{(\varepsilon-\mu)/kT} + a}$$

where

$$\begin{aligned} a &= +1 \quad \text{for fermions} \\ &= 0 \quad \text{for maxwellons} \\ &= -1 \quad \text{for bosons} \end{aligned}$$

and the exponential is much greater than unity. We then reexpress $n(\varepsilon)$ as

$$n(\varepsilon) = \frac{e^{(\mu-\varepsilon)/kT}}{1 + ae^{(\mu-\varepsilon)/kT}}$$

so that the exponential in the denominator is now small and we can thus perform a binomial expansion:

$$\begin{aligned} n(\varepsilon) &= e^{(\mu-\varepsilon)/kT}\{1 + ae^{(\mu-\varepsilon)/kT}\}^{-1} \\ &= e^{(\mu-\varepsilon)/kT}\{1 - ae^{(\mu-\varepsilon)/kT} + a^2 e^{2(\mu-\varepsilon)/kT} - \cdots\}. \end{aligned}$$

We shall use this series expansion to evaluate the "Fermi energy". Of course, the concept of the Fermi energy is peculiar to Fermi systems; it represents the highest filled energy state at $T = 0$. And this is entirely meaningless for bosons and classical particles, where the Pauli exclusion principle does not apply. Nevertheless, we shall adopt an "effective" Fermi energy through the expression of Eq. (2.25):

$$\varepsilon_F = \frac{\hbar^2}{2m}\left(\frac{6\pi^2}{\alpha}\frac{N}{V}\right)^{2/3}.$$

Thus in effect we are using ε_F simply as a characteristic energy that conveniently parameterises the number density N/V.

In order to find the chemical potential we must then invert the expression, from Eq. (2.34),

$$\varepsilon_F^{3/2} = \frac{3}{2}\int_0^\infty n(\varepsilon)\varepsilon^{1/2}\,d\varepsilon$$

which we shall do in terms of the above expansion.

$$\varepsilon_F^{3/2} = \frac{3}{2}\int_0^\infty \left\{ e^{(\mu-\varepsilon)/kT} - ae^{2(\mu-\varepsilon)/kT} + a^2 e^{3(\mu-\varepsilon)/kT} - \cdots \right\} \varepsilon^{1/2}\,d\varepsilon.$$

This can be integrated term by term to give

$$\varepsilon_F^{3/2} = (kT)^{3/2}\left\{ \frac{3}{4}\sqrt{\pi}e^{\mu/kT} - \frac{3}{8}a\sqrt{\frac{\pi}{2}}e^{2\mu/kT} + \frac{1}{4}a^2\sqrt{\frac{\pi}{3}}e^{3\mu/kT} \right.$$
$$\left. - \frac{3}{32}a^3\sqrt{\pi}e^{4\mu/kT} + \cdots \right\}.$$

We regard this as a power series in $e^{\mu/kT}$, known as the fugacity, which we can then invert to give

$$e^{\mu/kT} = \frac{4}{3\sqrt{\pi}}\left(\frac{\varepsilon_F}{kT}\right)^{3/2} + a\frac{4\sqrt{2}}{9\pi}\left(\frac{\varepsilon_F}{kT}\right)^3 + a^2\frac{16\left(9-4\sqrt{3}\right)}{243\pi^{3/2}}\left(\frac{\varepsilon_F}{kT}\right)^{9/2} + \cdots$$

In the classical case we take $a = 0$, whereupon

$$e^{\mu/kT} = \frac{4}{3\sqrt{\pi}}\left(\frac{\varepsilon_F}{kT}\right)^{3/2}$$

or

$$\mu_{classical} = kT\log\left[\frac{4}{3\sqrt{\pi}}\left(\frac{\varepsilon_F}{kT}\right)^{3/2}\right].$$

If we substitute for ε_F from Eq. (2.25) then we arrive at the previously obtained expression for the classical gas chemical potential in Eq. (2.20).

We can express the series for the fugacity as

$$e^{\mu/kT} = \frac{4}{3\sqrt{\pi}} \left(\frac{\varepsilon_F}{kT}\right)^{3/2} \left[1 + a\frac{1}{3}\sqrt{\frac{2}{\pi}}\left(\frac{\varepsilon_F}{kT}\right)^{3/2} + a^2\frac{4(9 - 4\sqrt{3})}{81\pi}\left(\frac{\varepsilon_F}{kT}\right)^3 + \cdots\right].$$

Then, upon taking the logarithm, we can express the high temperature/low density chemical potential in terms of its classical value plus a series of correction terms

$$\mu = \mu_{\text{classical}} + kT\left\{a\frac{1}{3}\sqrt{\frac{2}{\pi}}\left(\frac{\varepsilon_F}{kT}\right)^{3/2} + a^2\frac{(27 - 16\sqrt{3})}{81\pi}\left(\frac{\varepsilon_F}{kT}\right)^3 + \cdots\right\}.$$

(2.40)

In the Fermi case ($a = +1$) the chemical potential is above the classical value while in the Bose case ($a = -1$) the chemical potential falls below the classical value.

Observe that μ/ε_F may be written as a function of the single variable kT/ε_F. In particular, the classical/maxwellon chemical potential is

$$\frac{\mu}{\varepsilon_F} = \frac{kT}{\varepsilon_F}\log\left[\frac{4}{3\sqrt{\pi}}\left(\frac{\varepsilon_F}{kT}\right)^{3/2}\right].$$

Upon substituting for the Fermi energy, we obtain the classical ideal gas chemical potential as

$$\mu = -kT\ln\left[\alpha\left(\frac{mkT}{2\pi\hbar^2}\right)^{3/2}\frac{V}{N}\right]$$

in agreement with that calculated in Sec. 2.3.3.

Using a series procedure similar to that above (Appendix 3.4), we may find the internal energy of the three gases as

$$E = \frac{3}{2}NkT\left\{1 + a\sqrt{\frac{2}{\pi}}\frac{1}{6}\left(\frac{\varepsilon_F}{kT}\right)^{3/2} + a^2\cdots\right\}$$

(2.41)

where we recall that $a = -1$ for fermions, 0 for maxwellons and $+1$ for bosons. And then from the general relation, Eq. (2.10), $pV = 2E/3$ we obtain the equation of state for the gases as

$$pV = NkT\left\{1 + a\sqrt{\frac{2}{\pi}}\frac{1}{6}\left(\frac{\varepsilon_F}{kT}\right)^{3/2} + a^2\cdots\right\}.$$

(2.42)

Observe that the pressure of the fermion gas is increased over that of the classical gas, while the pressure of the boson gas is decreased.

Some tables and series expressions for the thermodynamic proper-ties of Fermi gases are given by Ebner and Fu[5] and references therein. There are some informative *Mathematica* notebooks by J. Kelly published on the Web.[6]

2.5. Ideal Bose Gas

There is no restriction on the number of bosons occupying a single state. This means that at low temperatures there will be a considerable occupa-tion of the low-energy states. This is in contrast to the Fermi case where the Pauli exclusion principle restricts the state occupation. Thus the behaviour of the Bose gas and the Fermi gas will be particularly different at low temperatures. Indeed the zero-temperature states are very different. The Fermi gas has all states up to the Fermi level occupied, with a resultant significant internal energy. Since there is no restriction on occupation, the ground state of the Bose gas will have all particles occupying the same lowest energy single-particle state. This macroscopic occupation of the lowest single-particle state is known as Bose–Einstein condensation; its possibility was predicted by Einstein in 1925. The early treatment of this phenomenon by London[7] is still one of the clearest.

2.5.1. *General procedure for treating the Bose gas*

The Bose–Einstein distribution is

$$n(\varepsilon) = \frac{1}{e^{(\varepsilon-\mu)/kT} - 1},$$

which gives the mean number of bosons in the state of energy ε in a system at temperature T and chemical potential μ. From this, the average value of a general function of energy $f(\varepsilon)$ is given by

$$\bar{f} = \sum_i f(\varepsilon_i)n(\varepsilon_i)$$

where the sum is taken over the single-particle energy states. Convention-ally, this sum is transformed to an integral over energy states using the density of states function $g(\varepsilon)$ as:

$$\bar{f} = \int_0^\infty f(\varepsilon)g(\varepsilon)n(\varepsilon)d\varepsilon,$$

where $g(\varepsilon)$ is given by Eq. (2.3):

$$g(\varepsilon) = \frac{V}{4\pi^2\hbar^3} (2m)^{3/2} \varepsilon^{1/2}.$$

The procedure, as described, will give the expression for $\bar{f}$ in terms of the system's characterising intensive parameters T and μ. But, as we have seen already, the chemical potential is best eliminated in terms of the total number of particles in the system. Thus the first task must be to consider the expression for the number of particles.

2.5.2. Number of particles — chemical potential

One is inclined to write the expression for the number of particles in the system by integrating over the number of particles in each quantum state:

$$N = \sum_i n_i \rightarrow \int_0^\infty g(\varepsilon) n(\varepsilon) d\varepsilon.$$

However there is a problem with this. We know that at low enough temperatures there will be a large number of particles in the ground state, of energy $\varepsilon = 0$. But the density of states $g(\varepsilon)$ is proportional to $\varepsilon^{1/2}$ (in three dimensions). This means that it gives *zero weight* to the ground state. There is an error introduced in transforming from a sum over states to an integral using the density of states. Ordinarily there will be no problem with the neglect of a single state. But if there is an appreciable occupation of this state then the error becomes serious. Since the $\varepsilon^{1/2}$ factor neglects completely the ground state we must add this "by hand" to the calculation for N. Thus we write (using a spin degeneracy factor α of unity):

$$N = N_0 + \frac{V}{4\pi^2\hbar^3} (2m)^{3/2} \int_0^\infty \frac{\varepsilon^{1/2}}{e^{(\varepsilon-\mu)/kT} - 1} d\varepsilon \qquad (2.43)$$

where N_0 is the number of particles in the ground state.

For the Bose–Einstein distribution, and since ε for the ground state is zero, it follows that

$$N_0 = \frac{1}{e^{-\mu/kT} - 1}.$$

If the ground state occupation is appreciable — above a few hundred, for instance, then the denominator will be small which means that the exponential will differ only slightly from unity. Then μ must be very small (and negative) so that

$$N_0 = \frac{1}{1 - \mu/kT + \cdots - 1} \sim -kT/\mu$$

or

$$\mu \sim -kT/N_0. \qquad (2.44)$$

We conclude that macroscopic occupation of the ground state is associated with a vanishingly small chemical potential.

2.5.3. *Low temperature behaviour of Bose gas*

At low temperatures, when N_0 is appreciable, then the chemical potential is very small and it can be ignored, compared with ε in the integral for the number of excited particles. In Problem 2.9 we see that this is so, typically, for $N_0 > 10^{15}$ or $N_0/N > 10^{-8}$. This is a very small *fraction* of the number of particles in the system. Then, within this approximation, the expression for N becomes

$$N = N_0 + \frac{V}{4\pi^2\hbar^3}(2m)^{3/2} \int_0^\infty \frac{\varepsilon^{1/2}}{e^{\varepsilon/kT} - 1}\, d\varepsilon.$$

The integral may be "tidied" by the substitution $x = \varepsilon/kT$ whereupon

$$N = N_0 + \frac{V}{4\pi^2\hbar^3}(2mkT)^{3/2} \int_0^\infty \frac{x^{1/2}}{e^x - 1}\, dx.$$

The integral in this expression is a pure number. Let us just denote it by $I_{1/2}$ for the moment, and consider its actual value later. Then the number of excited particles in the system is given by

$$N_{\text{ex}} = \frac{V}{4\pi^2\hbar^3}(2mkT)^{3/2} I_{1/2} \qquad (2.45)$$

which varies with temperature. There will be a certain temperature when N_{ex} will be equal to the total number of particles in the system N — at least according to this equation. In reality the expression will become invalid when N_0 becomes *too* small, but let us just use this as a mathematical definition of a characteristic temperature T_c, that is, let us define T_c by

$$N = \frac{V}{4\pi^2\hbar^3}(2mkT_c)^{3/2} I_{1/2}.$$

Then the expression for N becomes

$$N = N_0 + N \left(\frac{T}{T_c} \right)^{3/2} \tag{2.46}$$

which may be inverted to give the number of particles in the ground state as

$$N_0 = N \left\{ 1 - \left(\frac{T}{T_c} \right)^{3/2} \right\}. \tag{2.47}$$

This expression is valid for very low temperatures; clearly we must have $T < T_c$, but how close to T_c can one go? In fact one can go very close indeed since we require the *absolute* value of N_0 to be large. And as one has $N \sim 10^{23}$ it follows that the *fractional* value N_0/N will be small — perhaps 10^{-8}, while N_0 will still be *enormous* at $\sim 10^{15}$. So, as shown in Problem 2.9 we can use the above expression for N_0 for temperatures $T < T_c$ right up to within $\sim 10^{-8}\, T_c$ of T_c. Then we can identify the temperature T_c as a transition temperature, below which there will be a *macroscopic* occupation of the ground state. This is known as the *Bose–Einstein condensation* and it happens at temperature T_c. It is the only phase transition in nature that happens in the absence of any interactions.

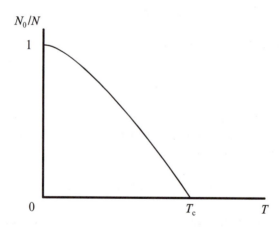

Fig. 2.7.   Ground state occupation for Bose gas.

The transition temperature is given by

$$T_c = \frac{2\pi^2 \hbar^2}{mk} \left\{ \frac{N}{2\pi I_{1/2} V} \right\}^{2/3} \tag{2.48}$$

where $I_{1/2}$ is the integral (different from the I integrals of the Fermi–Dirac case)

$$I_{1/2} = \int_0^\infty \frac{x^{1/2}}{e^x - 1} \, dx. \tag{2.49}$$

The integral may be expressed as

$$I_{1/2} = \Gamma(3/2)\zeta(3/2)$$

where $\Gamma(\ )$ is Euler's gamma function and $\zeta(\ )$ is the Riemann zeta function. These have values

$$\Gamma(3/2) = \sqrt{\pi}/2$$
$$\zeta(3/2) = 2.612\ldots$$

and from these we may write T_c as

$$T_c = \frac{2\pi\hbar^2}{mk} \left\{ \frac{N}{2.612V} \right\}^{2/3}$$
$$= 3.313 \frac{\hbar^2}{mk} \left\{ \frac{N}{V} \right\}^{2/3}. \tag{2.50}$$

2.5.4. *Thermal capacity of Bose gas — below T_c*

To find the thermal capacity of the Bose gas we must first obtain an expression for the internal energy. We shall do this all within the approximation of neglecting the chemical potential. In other words, our results will be valid for temperatures *below* the Bose–Einstein condensation temperature.

The internal energy is given by

$$E = \int_0^\infty \varepsilon g(\varepsilon) n(\varepsilon) d\varepsilon.$$

Here we do not need to worry about the ground state occupation since it contributes nothing to the internal energy. The expression for E is

$$E = \frac{V}{4\pi^2\hbar^3} (2m)^{3/2} \int_0^\infty \frac{\varepsilon^{3/2}}{e^{(\varepsilon-\mu)/kT} - 1} \, d\varepsilon,$$

and as in the previous case the integral may be "tidied" by the substitution $x = \varepsilon/kT$ whereupon

$$E = \frac{V}{4\pi^2 \hbar^3}(2m)^{3/2}(kT)^{5/2}\int_0^\infty \frac{x^{3/2}}{e^x - 1}\,dx.$$

Again, the integral in this expression is a pure number. By analogy with the previous case, let us just denote it by $I_{3/2}$ for the moment, and consider its actual value later. Then the internal energy of the assembly of Bosons is

$$E = \frac{V}{4\pi^2 \hbar^3}(2m)^{3/2}(kT)^{5/2}I_{3/2} \tag{2.51}$$

and upon differentiation the thermal capacity (at constant volume) is

$$C_V = \frac{5V}{8\pi^2 \hbar^3}k(2mkT)^{3/2}I_{3/2}. \tag{2.52}$$

But since the number of particles in the system is given by

$$N = \frac{V}{4\pi^2 \hbar^3}(2mkT_c)^{3/2}I_{1/2} \tag{2.53}$$

— in reality this is the definition of the transition temperature T_c — we may express the thermal capacity as

$$C_V = \frac{5}{2}Nk\frac{I_{3/2}}{I_{1/2}}\left(\frac{T}{T_c}\right)^{3/2}. \tag{2.54}$$

We already have the value of the integral $I_{1/2}$; it remains then, to find the value of the integral $I_{3/2}$. This is defined by

$$I_{3/2} = \int_0^\infty \frac{x^{3/2}}{e^x - 1}\,dx.$$

The integral may be expressed in terms of the gamma and zeta functions as

$$I_{3/2} = \Gamma(5/2)\zeta(5/2).$$

These have values

$$\Gamma(5/2) = 3\sqrt{\pi}/4$$
$$\zeta(3/2) = 1.342\ldots$$

so that below T_c the thermal capacity is given by

$$C_V = \frac{15}{4}Nk\left(\frac{T}{T_c}\right)^{3/2} \times \frac{1.342}{2.612}$$

$$= 1.926Nk\left(\frac{T}{T_c}\right)^{3/2}. \tag{2.55}$$

This should be compared with the classical value of $3Nk/2$. Above the transition temperature C_V will fall, gradually, to the classical value.

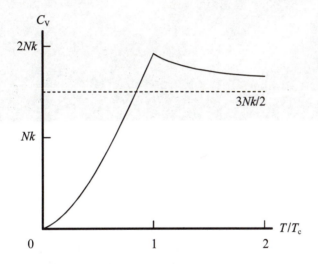

Fig. 2.8.   Thermal capacity of a Bose gas.

2.5.5.   *Comparison with superfluid ^{4}He and other systems*

In 1995 Bose–Einstein condensation was observed in rubidium vapour[8] and in lithium vapour.[9] An undemanding account of these discoveries is given in the article by Meacher and Ruprecht.[10] The density of these gases was very low so that the inter-particle interactions were very small. The condensation temperature was thus very low and temperatures in the nanokelvin range were required, achieved by laser confinement and laser cooling of the trapped vapour cloud.

The Bose–Einstein condensation was observed in rubidium vapour by imaging the density and momentum distribution of the atoms. Figure 2.9[11] shows the velocity distribution of rubidium atoms at three different

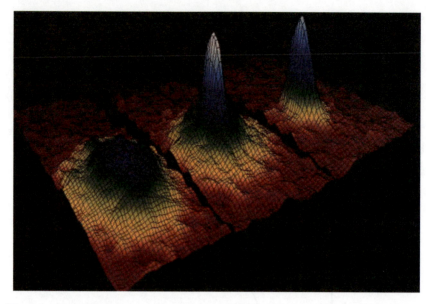

Fig. 2.9. Bose–Einstein condensation in rubidium vapour. Three velocity distributions of the rubidium atoms at three different temperatures.

temperatures. The left frame shows the velocity distribution at 400 nK, just before the appearance of the Bose–Einstein condensate. The centre frame is at 200 nK, just after the appearance of the condensate. And the right frame is at 50 nK, after further evaporation leaves a sample of nearly pure condensate. The area of view of each image is 200 μm by 270 μm. The appearance of the peak indicates the existence of a significant number of atoms in the ground state.

The interest in the two different materials was because rubidium has a remnant repulsive interaction while the remnant interaction in lithium is attractive; in both these cases the Bose–Einstein condensation was observed. However in both these cases the amount of material involved precluded the measurement of any thermodynamic properties. The case is different for liquid helium.

The common isotope of helium, ^{4}He, is found to go superfluid at a temperature of 2.172 K. In 1924, Kamerlingh Onnes observed that *something* strange happened to liquid ^{4}He at this temperature and in 1932 Keesom and Clausius observed a cusp in the heat capacity, indicative of a phase transition. Subsequently the low-temperature phase was observed to flow with no viscosity — superfluid behaviour. The calculated

Bose–Einstein condensation transition temperature for particles whose mass and volume corresponds to those of liquid ^{4}He is 3.13 K. Since these temperatures are similar it is possible that the superfluidity is related to the Bose–Einstein condensation. This was first suggested in 1938 by London[12] and by Tisza.[13] The possibility is supported by the fact that at these temperatures ^{3}He, a fermion, does not become superfluid; there is no such transition. The small difference in transition temperatures is ascribed to the interatomic interactions in liquid ^{4}He, which are neglected in the Bose gas model.

The following discussions will consider *interacting* bosons, so strictly this is a topic for the next chapter. However we shall place the discussion here as it follows naturally from our previous considerations.

2.5.6. *Two-fluid model of superfluid ^{4}He*

The two fluid model is a very successful thermodynamic/macroscopic model of superfluid helium. It was proposed by Tisza in 1938. The Bose–Einstein condensation gives a microscopic *motivation* for the model, but it cannot be regarded as a microscopic *justification* — as we shall see.

According to the two-fluid model the superfluid comprises a mixture of two interpenetrating fluids. One is the *normal* component and the other is the *superfluid* component. The density of the liquid is then expressed as the sum

$$\rho = \rho_{\mathrm{s}} + \rho_{\mathrm{n}}. \tag{2.56}$$

The normal component is assumed to have properties similar to the fluid above the transition. The superfluid component is assumed to have *zero entropy*. As the temperature is lowered below the transition temperature the superfluid fraction grows at the expense of the normal fraction. And at $T = 0$ the normal component has reduced to zero.

Originally, as proposed by Tisza, the connection with Bose–Einstein condensation was made by identifying the superfluid component with the macroscopically occupied ground state of the Bose gas. Since these particles are all in the same single-particle state there is only one microstate corresponding to this and its entropy is thus zero. However subsequently Landau proposed that the normal component should be associated with the elementary excitations of the system.

An experiment by Andronikashvili measured the normal component of superfluid helium using a stack of discs oscillating in the liquid.

Fig. 2.10.   Andronikashvili's experiment.

The normal component will move with the discs while the superfluid component will be decoupled. The resonant frequency of the disc assembly will depend on its effective mass. Thus the normal component can be measured through the resonant frequency of the discs.

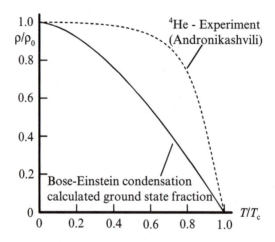

Fig. 2.11.   Ground state fraction.

The normal fraction, as measured by Andronikashvili, is compared with the fractional ground state occupation for the Bose–Einstein condensation in Fig. 2.11. There is a *qualitative* similarity, but the discrepancy is serious.

2.5.7. *Elementary excitations*

The difference between superfluid helium and the Bose–Einstein condensed fluid is made more apparent by measurements of thermal capacity.

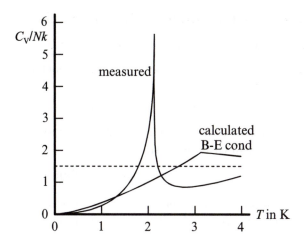

Fig. 2.12. Thermal capacity of liquid helium.

In Fig. 2.12, we show C_V as measured in liquid helium compared with that calculated for an ideal Bose gas. (Most books compare the calculated C_V with the observed thermal capacity under the liquid's saturated vapour pressure. This has a qualitatively different behaviour so the comparison is invalid.)

The low temperature behaviour of these two are different. The calculated C_V increases with temperature as $T^{3/2}$, but the observed thermal capacity varies as T^3. Now a T^3 thermal capacity indicates *phonon excitations*. Thus the interactions between the particles seem to quench the single-particle behaviour and leads to collective motion of the particles. This is supported by calculations of Bogoliubov,[14] who showed that even weak repulsive interactions change the elementary excitations of the system from free particle motion to phonon collective motion.

The conclusion is that the interactions cannot be regarded as causing a slight change to the properties of the Bose–Einstein condensate. Thus the two-fluid model cannot be built on the Bose–Einstein condensate idea. In reality the normal component comprises the elementary excitations in the system. The spectrum of elementary excitations is more complex than just phonons, but that is another story. A good account of the properties of superfluid helium-4 and their understanding is given in the book by Khalatnikov.[15] This book also contains translated reprints of the original papers by Landau on the subject.

2.6. Black Body Radiation — The Photon Gas

2.6.1. *Photons as quantised electromagnetic waves*

The harmonic oscillator has the very important property that its energy eigenvalues are equally spaced:

$$\varepsilon_n = n\hbar\omega + \text{zero point energy}.$$

You will have calculated the internal energy of the harmonic oscillator in Problem 1.16. We shall write this result here as

$$E = \frac{\hbar\omega}{e^{\hbar\omega/kT} - 1} + \text{zero point contribution}. \tag{2.57}$$

We shall, in the following sections, ignore the (constant) zero point energy contribution. This is acceptable since the zero of energy is, to a certain extent, arbitrary.

We now see that the expression for E can be reinterpreted in terms of the Bose distribution. The internal energy has the form

$$E = \bar{n}\hbar\omega + \text{zero point contribution} \tag{2.58}$$

where $\bar{n}$ is the Bose distribution, the mean number of bosons of energy $\hbar\omega$. But here, we observe in the Bose distribution that the chemical potential is zero.

The conclusion is that we can regard a harmonic oscillator of (angular) frequency ω as a collection of bosons of energy $\hbar\omega$, having zero chemical potential.

The fact that we can regard the harmonic oscillator as a collection of bosons is a consequence of the equal spacing of the oscillator energy levels. The vanishing of the chemical potential is due to the fact that the number of these bosons is not conserved.

We shall explore this by considering an isolated system of particles. If N is conserved then the number is determined — it is given and it will remain constant for the system. On the other hand if the number of particles is not conserved then one must determine the equilibrium number by maximising the entropy:

$$\left.\frac{\partial S}{\partial N}\right|_{E,V} = 0,$$

where we note that E and V are constant since the system is isolated. From the differential expression for the first law we see that

$$dS = \frac{1}{T}(dE + p\,dV - \mu\,dN).$$

So the entropy derivative is

$$\frac{\partial S}{\partial N}\bigg|_{E,V} = -\frac{\mu}{T}$$

and we conclude that the equilibrium condition for this system, at finite temperatures, is simply

$$\mu = 0.$$

In other words $\mu = 0$ for non-conserved particles.

2.6.2. *Photons in thermal equilibrium — black body radiation*

It is known that bodies glow and emit light when heated sufficiently. In 1859 Kirchhoff laid down a challenge: to derive the mathematical formula for the radiation spectrum. This challenge was one of the key factors in the development of quantum theory.

The spectrum of radiation emitted by a body depends on its temperature. The spectrum also depends on the nature of its surface — to a more or lesser extent. However Kirchhoff appreciated that there was an idealisation that could be considered, when the surface absorbs all the radiation that falls upon it. Such a surface appears to be black and Kirchhoff speculated on the spectrum of the radiation emitted by such a black body. He appreciated that the radiation was in thermal equilibrium with the body and that therefore the spectrum depended only on the temperature. The intensity of the spectrum was thus some function $U(\omega, T)$ and Kirchhoff's challenge was thus to find the form of this function.

The spectrum is then a property of the radiation and not of the body under consideration. Thus we can make a model system, an idealisation of the situation which retains the important features of the problem, but which is possible to solve.

Our model is simply a cavity, which is connected to the outside world by a small hole. We shall look through the hole at the spectrum of the radiation in the cavity.

We consider the electromagnetic waves to be in thermal equilibrium with the walls. The photons will have a distribution given by the Bose–Einstein formula, but with zero chemical potential. And we then calculate the properties of this photon gas by the methods of statistical mechanics.

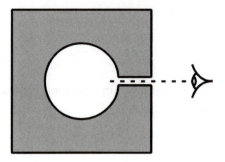

Fig. 2.13.   Radiation from a cavity.

2.6.3.   *Planck's formula*

In Sec. 2.1.5 we saw that the general expression for the energy density of states in three dimensions is given by Eq. (2.6):

$$g(\varepsilon) = \frac{V}{2\pi^2} k^2 \left/ \frac{d\varepsilon}{dk} \right.,$$

where k is to be eliminated in terms of ε. For photons there is a linear relation between ε and k:

$$\varepsilon = \hbar c k$$

where c is the speed of light. Thus the density of states is

$$g(\varepsilon) = \frac{1}{2} \frac{V \varepsilon^2}{c^3 \pi^2 \hbar^3}. \tag{2.59}$$

Now although photons have a spin $S = 1$, they have zero mass. Quantum-mechanically this means that only $S_z = -1$ and $+1$ orientations occur; $S_z = 0$ orientation is suppressed. Classically this is understood because electromagnetic waves propagate as transverse disturbances; there are no longitudinal waves. The photons' two quantum states correspond to the two polarisation states of the electromagnetic waves. This means that we must take a degeneracy factor α or photons to be 2. And so the internal energy for the photon gas in thermal equilibrium is given by

$$E = 2 \int_0^\infty \frac{g(\varepsilon)\varepsilon d\varepsilon}{e^{\varepsilon/kT} - 1}$$

$$= \frac{V}{c^3 \pi^2 \hbar^3} \int_0^\infty \frac{\varepsilon^3 d\varepsilon}{e^{\varepsilon/kT} - 1}. \tag{2.60}$$

Before we evaluate this integral we shall make connection with Kirchhoff's $U(\omega,T)$ introduced in Sec. 2.6.1. To this end, we shall change variables to $\omega = \varepsilon/\hbar$ so that

$$E = \frac{V\hbar}{c^3\pi^2} \int_0^\infty \frac{\omega^3 d\omega}{e^{\hbar\omega/kT} - 1}. \qquad (2.61)$$

Now the energy can be expressed in terms of the spectrum $U(\omega,T)$ by integrating over all frequencies

$$E = \int_0^\infty U(\omega, T)d\omega. \qquad (2.62)$$

From these two expressions we can then identify

$$U(\omega, T) = \frac{V\hbar}{c^3\pi^2} \frac{\omega^3}{e^{\hbar\omega/kT} - 1}. \qquad (2.63)$$

This is Planck's formula for black body radiation. Some examples are shown in Fig. 2.14, but plotted as a function of wavelength, which is more popular with spectroscopists.

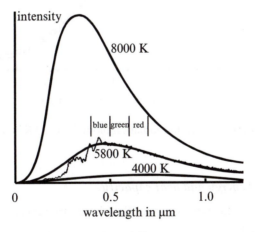

Fig. 2.14. Black body radiation at three different temperatures; the wiggly line is the spectrum from the sun.

The spectrum from the sun indicates that its temperature is about 5800 K. It is also of interest to note that the peak of the sun's spectrum corresponds to the visible spectrum, that is the region of greatest sensitivity of the human eye.

A remarkable example of black body radiation is the spectrum of electromagnetic radiation arriving from outer space. It is found that when looking into space with radio telescopes, a uniform background electromagnetic "noise" is seen. The spectrum of this is found to fit the black body curve — for a temperature of approximately 2.7 K. The conclusion is that the equilibrium temperature of the universe is 2.7 K, which is understood as the remnant "glow" from the Big Bang. The data shown in Fig. 2.15 comes from the COBE satellite and it fits the Planck black body curve for a temperature of 2.74 K. The quality of the fit of the data to the theoretical curve is quite remarkable.

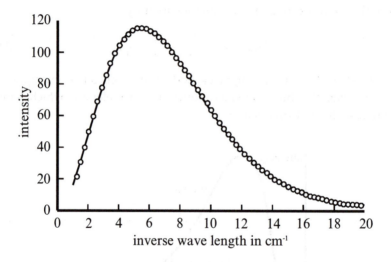

Fig. 2.15.   Cosmic background radiation plotted on 2.74 K black body curve.

2.6.4. *Internal energy and heat capacity*

We now integrate the expression for the photon free energy in the previous section, Eq. (2.60)

$$E = \frac{V}{c^3 \pi^2 \hbar^3} \int_0^\infty \frac{\varepsilon^3 d\varepsilon}{e^{\varepsilon/kT} - 1}.$$

By changing the variable of integration to $x = \hbar\omega/kT$ the integral becomes a dimensionless number

$$E = \frac{V\hbar}{\pi^2 c^3} \left(\frac{kT}{\hbar}\right)^4 \int_0^\infty \frac{x^3 dx}{e^x - 1}.$$

The integral may be evaluated numerically or it may be found in terms of the gamma function to give

$$\int_0^\infty \frac{x^3 dx}{e^x - 1} = \frac{\pi^4}{15}$$

so that the internal energy is

$$E = \frac{\pi^2 V \hbar}{15c^3} \left(\frac{kT}{\hbar}\right)^4. \tag{2.64}$$

The heat capacity is found by differentiating the internal energy

$$C_V = \left. \frac{\partial E}{\partial T} \right|_V,$$

giving

$$C_V = \frac{4\pi^2 V k^4}{15 \hbar^3 c^3} T^3. \tag{2.65}$$

We see that the photon heat capacity is proportional to T^3. Also note that the heat capacity goes to zero as $T \to 0$ in conformity with the Third Law.

2.6.5. *Black body radiation in one dimension*

We shall see that by considering black body radiation in one dimension, we can make an important connection with electrical noise, a topic that will be treated in Chapter 5. We may regard a waveguide as a one-dimensional black body cavity, and to start with we shall imagine this waveguide to connect black body cavities at either end. We imagine the cavities and the waveguide to be at a common temperature T and we consider the equilibrium thermal radiation in the waveguide.

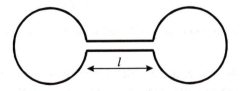

Fig. 2.16.   Waveguide joining two black body cavities.

In a one dimension cavity of length l the energy density of states is given by Eq. (2.8):

$$g(\varepsilon) = \frac{l}{\pi} \bigg/ \frac{d\varepsilon}{dk},$$

where $\varepsilon = c\hbar k$ for photons. Thus, the density of states is

$$g(\varepsilon) = \frac{l}{\pi c \hbar}, \tag{2.66}$$

independent of energy. The internal energy is then found by integrating:

$$E = 2 \int_0^\infty \frac{\varepsilon g(\varepsilon) d\varepsilon}{e^{\varepsilon/kT} - 1}$$

$$= \frac{2l}{\pi c \hbar} \int_0^\infty \frac{\varepsilon d\varepsilon}{e^{\varepsilon/kT} - 1}.$$

In the three-dimensional case, we made connection with the radiation spectrum by changing variables to $\omega = \varepsilon/\hbar$. However in the electrical case it is conventional to work in terms of the frequency f, so we change variables to $f = \varepsilon/2\pi\hbar = \varepsilon/h$ so that

$$E = \frac{8\pi l \hbar}{c} \int_0^\infty \frac{f df}{e^{2\pi \hbar f/kT} - 1}. \tag{2.67}$$

This enables us to identify the radiation spectrum $U(f, T)$ as

$$U(f, T) = \frac{4l}{c} \frac{hf}{e^{hf/kT} - 1}. \tag{2.68}$$

From the electrical point of view an infinitely long transmission line is equivalent to a finite line terminated with its characteristic resistance.

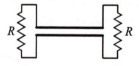

Fig. 2.17.   Transmission line terminated with its characteristic resistance.

In this case the spectrum of radiation in the transmission line will be the same. Thus the resistors may be regarded as emitting and absorbing the radiation. The time for the radiation to propagate along the waveguide is l/c. Thus the rate of energy flow, the power, in a frequency range Δf is given by

$$P_{\Delta f} = 4\frac{hf}{e^{hf/kT} - 1}\Delta f. \tag{2.69}$$

This may be interpreted as a random current or voltage in the resistors, in a frequency range Δf, of mean square value

$$\langle V^2 \rangle_{\Delta f} = 4R\frac{hf}{e^{hf/kT} - 1}\Delta f$$

$$\langle I^2 \rangle_{\Delta f} = \frac{4}{R}\frac{hf}{e^{hf/kT} - 1}\Delta f. \tag{2.70}$$

In the high-temperature/equipartition limit these expressions become

$$\langle V^2 \rangle_{\Delta f} = 4kTR\Delta f$$

$$\langle I^2 \rangle_{\Delta f} = \frac{4}{R}kT\Delta f, \tag{2.71}$$

independent of h and frequency. These random voltage/current fluctuations are a consequence of the thermal motion of the current-carriers in the resistors. As we shall see in Chapter 5, this phenomenon was predicted by Einstein in 1906, observed by Johnson in 1927. The spectrum of the fluctuations was explained theoretically by Nyquist in the same year.

2.7. Ideal Paramagnet

2.7.1. *Partition function and free energy*

Our treatment of the ferromagnet and the paramagnet–ferromagnet transition in Chapter 4 will be based upon an extension of the conventional model for the paramagnet — that is, an assembly of non-interacting, stationary magnetic moments. For this reason we shall present, in the following few sections, a summary of the properties of the ideal paramagnet. For convenience we shall consider magnetic moments of spin $1/2$. This will correspond to electrons, and the treatment is simplified for that case. In a magnetic field **B** a magnetic moment μ will have an energy $-\mu \cdot \mathbf{B}$. Now a spin $1/2$ may point either parallel or antiparallel to the applied magnetic field. Then the energy of these states will be

$$\varepsilon_\uparrow = -\mu B = -\gamma \hbar B/2$$
$$\varepsilon_\downarrow = +\mu B = +\gamma \hbar B/2$$

where γ is the magnetogyric (gyromagnetic) ratio of the moments. We denote the energy μB by ε, then the partition function Z for the entire system is

$$Z = z^N$$

where z is the partition function for a single one of the N spins. This is the expression for a collection of *distinguishable* identical subsystems. Although the spins themselves are certainly indistinguishable, we are considering a solid assembly where the subsystems can be distinguished by their positions. Now z is the sum over the two states

$$z = \sum_i e^{-\varepsilon_i/kT} = e^{\varepsilon/kT} + e^{-\varepsilon/kT} = 2\cosh(\varepsilon/kT)$$

where $\varepsilon = \mu B$. Then

$$Z = \{2\cosh(\varepsilon/kT)\}^N \tag{2.72}$$

and the connection with thermodynamics is made through the expression for the (magnetic) free energy

$$F = -kT \ln Z.$$

Thus we have

$$F = -NkT \ln(2\cosh\{\varepsilon/kT\}). \tag{2.73}$$

2.7.2. *Thermodynamic properties*

The various thermodynamic properties are found through differentiation. The differential of F, in the magnetic case, is

$$dF = -SdT - MdB. \tag{2.74}$$

The magnetic moment M is given by

$$M = \left.\frac{\partial F}{\partial B}\right|_T = N\mu \tanh\left(\frac{\varepsilon}{kT}\right)$$

or

$$M = N\mu \tanh\left(\frac{\mu B}{kT}\right); \tag{2.75}$$

this is the equation of state for the paramagnet.

Figure 2.18 shows the magnetisation as a function of $\varepsilon/kT = \mu B/kT$. At small fields B we have linear behaviour but saturation sets in at higher fields/lower temperatures when the dipoles are all aligned along the field so that no more magnetisation can be created.

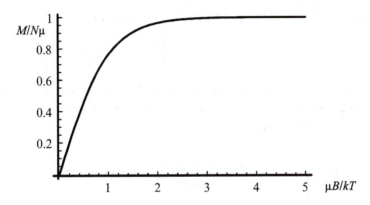

Fig. 2.18. Magnetisation of a paramagnet.

In the low field/high temperature limit the tanh may be expanded to first order: $\tanh x \sim x$. Then

$$M = N\mu^2 \frac{B}{kT}.$$

In this limit we have M proportional to B and inversely proportional to T. Then the magnetic susceptibility $\chi = \mu_0 M/VB$ is

$$\chi = \mu_0 \frac{N}{V} \frac{\mu^2}{kT}. \tag{2.76}$$

This result is known as Curie's law. One conventionally writes the magnetisation as

$$M = C\frac{B}{T} \tag{2.77}$$

where C is known as the Curie constant.

Observe that Curie's law is like the ideal gas equation of state:

$$M = \text{const} \times \frac{B}{T}$$

is like

$$\frac{1}{V} = \text{const} \times \frac{p}{T}.$$

In this respect we may regard Curie's law as the equation of state of the ideal paramagnet.

The other variable we find from dF is the entropy:

$$S = -\left.\frac{\partial F}{\partial T}\right|_{B} = Nk\left[\ln\left\{2\cosh\left(\frac{\varepsilon}{kT}\right)\right\} - \frac{\varepsilon}{kT}\tanh\left(\frac{\varepsilon}{kT}\right)\right]. \qquad (2.78)$$

This is shown as a function of temperature (kT/B). At high temperatures the entropy approaches $k\ln 2$ per particle as each moment has a choice of two states (two orientations).

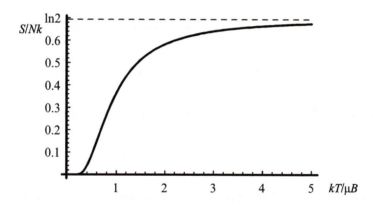

Fig. 2.19.   Entropy of a paramagnet.

We observe that the entropy tends to zero as T/B tends to zero. When the field B is finite this indicates behaviour in accord with the Third Law. The case of zero external fields must be treated carefully, as discussed in Sec. 1.7.6. The reality is that there will always be internal magnetic fields present: those produced by the dipoles themselves. Thus the Third Law is safe.

The internal energy is found as

$$E = F + TS = -N\mu B\tanh\left(\frac{\mu B}{kT}\right), \qquad (2.79)$$

which, we see, corresponds to $-MB$.

The heat capacity is given by:

$$C_B = \left.\frac{\partial E}{\partial T}\right|_B = Nk\left(\frac{\mu B}{kT}\right)^2 \text{sech}^2\left(\frac{\mu B}{kT}\right). \qquad (2.80)$$

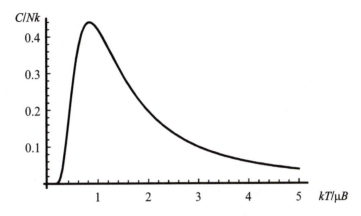

Fig. 2.20. Thermal capacity of a paramagnet.

The thermal capacity of the paramagnet has a characteristic peak at $kT = \mu B$. The peak is a consequence of the magnetic energy being bounded from above (unlike mechanical energy). This is known as the Schottky peak.

We finish this section by considering the equilibrium fluctuations in the magnetisation of a paramagnet. The magnetic moment of each spin $1/2$ particle is plus or minus μ. The square of the magnetic moment of each particle is then μ^2, so that the mean square magnetisation for the assembly of N particles is then

$$\langle M^2 \rangle = N\mu^2.$$

This may be regarded as the mean square of the fluctuations in the magnetisation from its mean value of zero. This may be related to the Curie law susceptibility in Eq. (2.76), thus

$$\chi = \mu_0 \frac{1}{VkT} \langle M^2 \rangle. \qquad (2.81)$$

Here we see how the magnetisation response function, the magnetic susceptibility, is related to the equilibrium fluctuations in the magnetisation. This is analogous to our result of Sec. 1.4.5 where we saw that the heat capacity (energy response function) was related to the equilibrium energy fluctuations. Indeed these are both special cases of the *fluctuation–dissipation* theorem, which we will encounter in Chapter 5.

2.7.3. *Negative temperatures*

The occupation of the up state and the down state are given by the appropriate Boltzmann factors. Since, as we have seen the energies of the two states are

$$\varepsilon_\uparrow = -\mu B$$
$$\varepsilon_\downarrow = +\mu B,$$

it follows that since

$$p_\uparrow = e^{-\varepsilon_\uparrow/kT}/z, \quad p_\downarrow = e^{-\varepsilon_\downarrow/kT}/z,$$

then the number of spins in the up and the down states are

$$N_\uparrow = Ne^{\mu B/kT}/z, \quad N_\downarrow = Ne^{-\mu B/kT}/z.$$

And then we see that the ratio of occupation of the up and down states is

$$\frac{N_\uparrow}{N_\downarrow} = e^{2\mu B/kT}. \tag{2.82}$$

Thus ordinarily there will be more spins in the up state than the down state; this follows since the up state has a lower energy than the down state.

It is clear that by pumping energy into this system, we will flip spins from the lower energy "up" state to the higher energy "down" state. And we can conceive of pumping energy, in this way until the populations of the two states becomes equalised. Moreover, we could continue to pump even more energy into the system so that there are more spins in the (higher energy) down state than the (lower energy) up state. This would be a strange configuration for our system, but in the interests of simplicity we could still apply Eq. (2.82) to describe it. In particular, we could use this equation to specify the temperature of our system. Then we would reach the following conclusions:

1. The "normal" situation, where the occupation of the lower energy state is higher than the occupation of the upper energy state corresponds to a situation where the temperature is a positive quantity.

2. When the populations of the two states are equalised, this corresponds to an *infinite* temperature.
3. When the occupation of the higher energy state is higher than the occupation of the lower energy state, the system would have a *negative* temperature.

Things become clearer when we plot the resultant temperature against the population ratio. We observe that there is a linear relation between *inverse* temperature and the logarithm of the population ratios. So in this context the inverse temperature is a more natural variable to use. In the figure we see that starting at large occupation of the up state corresponds to a low (positive) temperature — far right-hand side. Then feeding energy in, so that the up state occupation decreases, the temperature increases — moving to the left. The temperature diverges to plus infinity as the populations equalise. And then with greater occupation of the down state, the temperature returns from minus infinity and then it increases towards zero from the negative side.

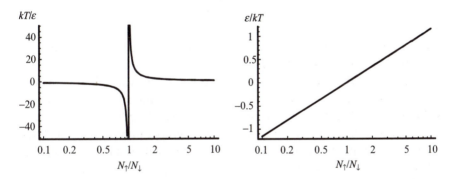

Fig. 2.21.   Temperature and inverse temperature as a function of population ratio.

It may be argued that the discussion justifying the concept of negative temperatures is somewhat artificial. It relies upon inversion of the Boltzmann factor expression for the population probabilities. Now the Boltzmann factor $\exp -\varepsilon/kT$ gives the population probabilities for a system in thermal equilibrium at a temperature T. But to what extent can we regard the states discussed in the section above as being in thermal equilibrium?

We were able to sidestep this question in the introductory discussion through restricting the consideration to spin $1/2$ particles where each has the choice of two orientations. Then any value for $N_\uparrow$ and $N_\downarrow$ will give a temperature value through Eq. (2.76). The situation is different, however, for spins $S > 1/2$. If there are more that two possible spin states then only specific numbers of particles occupying the different states will correspond to the Boltzmann factor expression and thus to states of thermal equilibrium.

As an example, consider the case of $S = 1$, where there are three spin orientations: $S_z = -1, 0$ and $+1$. The energies of these states, in a magnetic field B, are

$$\varepsilon_1 = -\mu B$$
$$\varepsilon_0 = 0$$
$$\varepsilon_{-1} = +\mu B$$

so that the populations of the three states will be

$$N_1 = N e^{\mu B/kT}/z$$
$$N_0 = N/z$$
$$N_{-1} = N e^{-\mu B/kT}/z.$$

And only when the populations obey a relation such as this, can one ascribe a temperature to the system.

If we use the ratio N_1/N_{-1} as the independent variable (by analogy with the spin $1/2$ case) then in this case we have the supplementary condition $N_0 = \sqrt{N_1 N_{-1}}$. Only when this is satisfied, will the system be in thermal equilibrium and we can ascribe a temperature to it. In real systems the (small) interactions between the spins will be such as to establish this thermal equilibrium. We will return to this point at the end of the next section.

2.7.4. *Thermodynamics of negative temperatures*

The thermodynamic definition of temperature, from the First Law, is given by

$$\frac{1}{T} = \frac{\partial S}{\partial E}.$$

Now in "normal" systems, the entropy is a monotonically increasing function of the energy. In this case, the derivative will be positive and so negative temperatures will not arise. We see that to have a negative temperature the entropy must be a *decreasing* function of the energy. This is possible with a spin system.

At the lowest energy all spins will be in their lowest energy state; all spins in the same state then means low (zero) entropy. Then as the energy increases, an increasing number of spins will be promoted to the higher state. Here the disorder is increasing and so the entropy is decreasing. With sufficient energy the populations will become equalised; this corresponds to the maximum entropy possible. And then as the energy continues to increase, an increasing number of spins will find themselves in the higher energy state. Finally, at the maximal energy, all spins will be in the higher energy state and the entropy once again will become zero. This behaviour will now be calculated.

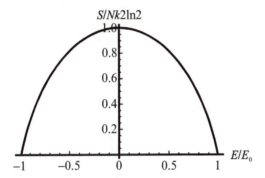

Fig. 2.22.   Entropy of a spin $1/2$ paramagnet as a function of energy.

The entropy is most conveniently expressed in terms of the Gibbs expression

$$S = -Nk \sum_j p_j \ln p_j$$

where the p_j are the probabilities of the single-particle states. Restricting our discussion to the spin $1/2$ case, there are two states to consider:

$$S = -Nk \{ p_\uparrow \ln p_\uparrow + p_\downarrow \ln p_\downarrow \}.$$

These probabilities may be expressed in terms of the magnetisation as

$$p_\uparrow = \frac{1 + M/N\mu}{2}, \quad p_\downarrow = \frac{1 - M/N\mu}{2},$$

or, since the energy is given by $E = -MB$,

$$p_\uparrow = \frac{1 - E/E_0}{2}, \quad p_\downarrow = \frac{1 + E/E_0}{2}$$

where E_0 is the energy corresponding to the saturation magnetisation: $E_0 = N\mu B$. Thus the entropy is given by

$$S = \frac{Nk}{2}\left\{2\ln 2 - \left(1 + \frac{E}{E_0}\right)\ln\left(1 + \frac{E}{E_0}\right) - \left(1 - \frac{E}{E_0}\right)\ln\left(1 - \frac{E}{E_0}\right)\right\}.$$

This is shown in Fig. 2.22.

It should be appreciated that a fundamental requirement for the existence of negative temperatures is that the system's energy must be bounded from above.

In the previous section we showed the heat capacity of the paramagnet for positive temperatures. The same expression holds for negative temperatures. Thus, we obtain the heat capacity shown in Fig. 2.23. Note that the heat capacity remains positive at negative temperatures. This ensures stability of the system.

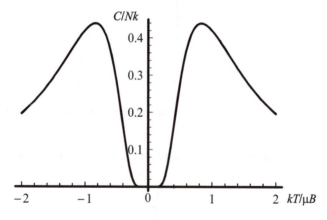

Fig. 2.23.   Heat capacity of a paramagnet at positive and negative temperatures.

We finish this section with some general comments about negative temperature states. Since a fundamental requirement for states of negative temperature is that the energy must be bounded from above, this means that in any real system the degree of freedom with such bounded energy states must be thermally isolated from the normal (unbounded) degees of freedom. Then the negative temperature applies to the subsystem and it is, in reality, a quasi-equilibrium state.

Such quasi-equilibrium states are of interest only to the extent that they are reasonably long-lived. And they are only meaningful to the extent that the quasi-equilibrium state is established on a timescale much shorter than this. The first observations of negative temperatures were in the nuclear spins of lithium in LiF crystals.[16] There the relaxation time for the interaction between the spin degrees of freedom and the normal degrees of freedom was of the order of five minutes while the internal equilibrium of the Li nuclear spins was established in less than 10^{-5} second.

Problems

2.1. In Sec. 2.1 we saw that the density of free-particle states for a three-dimensional volume V was shown to be

$$g(\varepsilon) = \frac{1}{4} \frac{V}{\pi^2 \hbar^3} (2m)^{3/2} \varepsilon^{1/2}.$$

This followed from counting the number of states in the octant of radius

$$R = \sqrt{n_x^2 + n_y^2 + n_z^2}.$$

By similar arguments show that in two dimensions, by counting the number of states in the quadrant of radius

$$R = \sqrt{n_x^2 + n_y^2},$$

the density of states is given by

$$g(\varepsilon) = \frac{mA}{2\pi\hbar^2}$$

where A is the area. Note in two dimensions the density of states is independent of energy.

And similarly, show that in one dimension the density of states is

$$g(\varepsilon) = \frac{L}{\pi\hbar} \left(\frac{m}{2} \right)^{1/2} \varepsilon^{-1/2}.$$

2.2. In Secs. 2.3.1 and 2.3.2 the ideal gas partition function was calculated quantum-mechanically and classically. Although the calculations were quite different, they both resulted in (different) Gaussian integrals. By writing the Gaussian integral of the classical case as

$$\int_{-\infty}^{\infty} dx \int_{-\infty}^{\infty} dy \int_{-\infty}^{\infty} dz \, e^{-(x^2+y^2+z^2)}$$

and transforming to spherical polar coordinates, you can perform the integration over θ and φ trivially. Show that the remaining integral can be reduced to that of the quantum case.

2.3. The Sakur–Tetrode equation, discussed in Sec. 2.3.3,

$$S = Nk \ln V - Nk \ln N + \frac{3}{2} Nk \ln T + Nks_0$$

is often interpreted as indicating different contributions to the entropy: the volume contribution is in the first term, the number contribution in the second term and the temperature contribution in the third term. Show that such an identification is fallacious, by demonstrating that the various contributions depend on the choice of units adopted — even though the total sum is independent. Discuss the origin of the fallacy.

2.4. Show that the Fermi energy for a two-dimensional gas of fermions is

$$\varepsilon_F = \frac{2\pi \hbar^2}{\alpha m} \frac{N}{A}$$

where A is the area of the system.

2.5. Show that the chemical potential of a two-dimensional gas of fermions may be expressed analytically as

$$\mu = kT \ln \{ e^{\varepsilon_F/kT} - 1 \}.$$

2.6. Calculate the low temperature chemical potential of a two-dimensional gas of fermions by the Sommerfeld expansion method of Sec. 2.4.3. Observe that the temperature series expansion terminates. Compare this result with the exact result of the previous question. Discuss the difference between the two results.

2.7. The general formula for the Fermi integrals I_n of Sec. 2.4.3 was quoted as

$$I_n = \int_{-\infty}^{\infty} \frac{e^x}{(e^x+1)^2} x^n dx$$

$$= (2 - 2^{2-n}) \zeta(n) n!$$

Derive this result. (You might find the discussion in Landau and Lifshitz, Statistical Physics, helpful.)

2.8. Obtain the chemical potential μ, the internal energy E and the heat capacity C_v for system with general density of states $g(\varepsilon)$ as in Sec. 2.4.4. That is, show that these are given in terms of the behaviour of the density of states at the Fermi surface.

2.9. Consider the Bose gas at low temperatures. You saw in Secs. 2.5.2 and 2.5.3 that when the occupation of the ground state is appreciable then the chemical potential μ is very small and it may be ignored, compared with ε in the integral for the number of excited states.

Show that when the ground state occupation N_0 is appreciable then μ may be approximated by

$$\mu \sim -kT/N_0.$$

Now consider the more stringent requirement that μ may be neglected in comparison with ε in the integral for the number of excited states. This will be satisfied if μ is much less than the energy ε_1 of the first excited state. The expression for ε_1 is

$$\varepsilon_1 \sim \frac{\pi^2 \hbar^2}{2m V^{2/3}}.$$

Where does this expression come from?

Show that the condition $\mu \ll \varepsilon_1$ is satisfied when $N \gg 10^{15}$ (approximately) when considering $1\,\text{cm}^3$ of ^4He (molar volume $27\,\text{cm}^3$) at a temperature of about 1 K.

Thus, show that the expression

$$N_0 = N \left\{ 1 - \left(\frac{T}{T_c} \right)^{3/2} \right\}$$

is then valid to temperatures below T_c right up to within $\sim 10^{-8} T_c$ of the critical temperature.

2.10. Liquid ^4He has a molar volume at saturated vapour pressure of $27\,\text{cm}^3$. Treating the liquid as an ideal gas of bosons, find the temperature at which Bose–Einstein condensation will occur. How will this temperature change as the pressure on the fluid is increased?

2.11. The superfluid transition temperature of liquid helium decreases with increasing pressure. Very approximately $\partial T_c/\partial p \sim -0.015\ \text{K bar}^{-1}$. How does this compare with the behaviour predicted from the Bose–Einstein condensation?

2.12. Show that below the transition temperature the entropy of a Bose gas is given by

$$S = \frac{5}{3} Nk \frac{I_{3/2}}{I_{1/2}} \left(\frac{T}{T_c} \right)^{3/2}.$$

Since the number of excited particles is given by

$$N_{\text{ex}} = N \left(\frac{T}{T_c} \right)^{3/2},$$

show that the entropy per excited particle is given by

$$\frac{S}{N_{\text{ex}}} = \frac{5}{3} \frac{I_{3/2}}{I_{1/2}} k \approx 1.28 \, k.$$

Discuss the connection between this result and the two-fluid model of superfluid ^{4}He.

2.13. Show that the Bose–Einstein transition temperature of a gas of bosons and the Fermi temperature for a gas of "similar" fermions are of comparable magnitude. Discuss why this should be.

2.14. In Sec. 2.6 we studied a paramagnetic *solid*: a collection of essentially *distinguishable* magnetic moments. If we were to consider a (classical) gas of indistinguishable magnetic moments, how would the partition function be modified? What would the observable consequences of this modification?

2.15. Show, using arguments similar to those in Sec. 2.1.3, that the energy levels of an ultra-relativistic or a massless particle with energy-momentum relation $E = cp$ are given by

$$\varepsilon = \frac{c\pi\hbar}{V^{1/3}} \left(n_x^2 + n_y^2 + n_z^2 \right)^{1/2}.$$

Hence show that the pressure of a gas of such particles is one third of the (internal) energy density.

2.16. Evaluate the Fermi temperature for liquid ^{3}He, assuming it to be a Fermi "gas". Its molar volume is $36 \, \text{cm}^3$. Calculate the de Broglie wavelength at $T = T_F$ and show that it is comparable to the interparticle spacing as expected.

2.17. In Problem 2.1 we found the expression for the energy density of states $g(\varepsilon)$ for a gas of fermions confined to two dimensions and we saw that it was independent of energy. What surface density of electrons is necessary in order that $T_F = 100 \, \text{mK}$? Show that, for a given area, the low temperature heat capacity is independent of the number of electrons.

2.18. Use the Sommerfeld expansion method of Sec. 2.4.3 to show that the Fermi–Dirac distribution function may be approximated, at low temperatures, by

$$\frac{1}{e^{(\varepsilon - \mu)/kT} + 1} \sim \Theta(\mu - \varepsilon) - \frac{\pi^2}{6}(kT)^2 \delta'(\varepsilon - \mu) + \cdots$$

where Θ is the unit step function and δ' is the first derivative of the Dirac delta function.

Can you write down the general term of the series?

References

[1] B. I. Bleaney and B. Bleaney, *Electricity and Magnetism*, 3rd ed. (Oxford University Press, 1976).

[2] A. Sommerfeld, *Z. Physik.* **47** (1928) 1.

[3] L. D. Landau and E. M. Lifshitz, *Statistical Physics* (Pergamon Press, 1980).

[4] F. Reif, *Fundamentals of Statistical and Thermal Physics* (McGraw-Hill, 1965).

[5] C. Ebner and Hui-Hsing Fu, *J. Low Temp. Phys.* **16** (1974) 43–50.

[6] J. Kelly, http://www.physics.umd.edu/courses/CourseWare/Statistical-Physics/

[7] F. London, *Superfluids, Vol II: Macroscopic Theory of Superfluid Helium* (John Wiley, 1954); Reprinted (Dover, 1964).

[8] M. H. Anderson, J. R. Ensher, M. R. Matthews, C. E. Weiman and E. A. Cornell, *Science* **269** (1995) 198.

[9] C. C. Bradley, C. A. Sackett, J. J. Tollett and R. G. Hulet, *Phys. Rev. Lett.* **75** (1995) 1687.

[10] D. Meacher and P. Ruprecht, *Physics World* **8** (1995) 21.

[11] E. Cornell, Very cold indeed: The Nanokelvin physics of Bose–Einstein condensation, *J. Res. Natl. Inst. Stand. Technol.* **101** (1996) 419.

[12] F. London, *Nature* **141** (1938) 643.

[13] L. Tisza, *Nature* **141** (1938) 913.

[14] N. Bogoliubov, *J. Phys.* (USSR), **11** (1947) 23.

[15] I. M. Khalatnikov, *An Introduction to the Theory of Superfluidity* (Benjamin, 1965).

[16] E. M. Purcell and R. V. Pound, *Phys. Rev.* **81** (1951) 279.

CHAPTER THREE

NON-IDEAL GASES

This chapter is devoted to considering systems where the interactions between particles can no longer be ignored. We note that in the previous chapter we did indeed consider, albeit briefly, the effects of interactions in fermion and in boson gases. This chapter is concerned more with a systematic treatment of interatomic interactions. Here the quantum aspect is a complication and most of the discussions will thus take place within the context of a classical description.

3.1. Statistical Mechanics

3.1.1. *The partition function*

We are now considering gases where the interactions between the particles cannot be ignored. Our starting point is that everything can be found from the partition function. We will work, initially, in the classical framework where the energy function of the system is

$$H\left(p_i, q_i\right) = \sum_i \frac{p_i^2}{2m} + \sum_{i<j} U\left(q_i, q_j\right).$$

Because of the interaction term $U\left(q_i, q_j\right)$ the partition function can no longer be factorised into the product of single-particle partition functions. The many-body partition function is

$$Z = \frac{1}{N! h^{3N}} \int e^{-\left(\sum_i \frac{p_i^2}{2m} + \sum_{i<j} U(q_i, q_j)\right) \Big/ kT} \mathrm{d}^{3N}p\, \mathrm{d}^{3N}q$$

where the factor $1/N!$ is used to account for the particles being indistinguishable.

While the partition function cannot be factorised into the product of single-particle partition functions, we can factor out the partition function for the non-interacting case since the energy is a sum of a momentum-dependent term (kinetic energy) and a coordinate-dependent term (potential energy). The non-interacting partition function is

$$Z_{id} = \frac{V^N}{N!h^{3N}} \int e^{-\sum_i \frac{p_i^2}{2mkT}} \, d^{3N}p$$

where the V factor comes from the integration over q_i. Thus the interacting partition function is

$$Z = Z_{id} \frac{1}{V^N} \int e^{-\left(\sum_{i<j} U(q_i, q_j)\right)/kT} \, d^{3N}q.$$

The "correction term" is referred to as the configuration integral. We denote this by Q_N

$$Q_N = \frac{1}{V^N} \int e^{-\left(\sum_{i<j} U(q_i, q_j)\right)/kT} \, d^{3N}q.$$

Different authors have different prefactors such as V or $N!$, but that is not important. The partition function for the interacting system is then

$$Z = \frac{1}{N!} \left(\frac{V}{\Lambda^3}\right)^N Q_N$$

and the attention now focuses on evaluation/approximation of the configuration integral Q_N.

3.1.2. *Cluster expansion*

We need a "small quantity" in terms of which to perform an expansion. To this end, we define

$$f_{ij} = e^{-U(q_i, q_j)/kT} - 1$$

so that f_{ij} is only appreciable when the particles are close together. In such terms the configuration integral is

$$Q_N = \frac{1}{V^N} \int \prod_{i<j} (1 + f_{ij}) d^{3N}q_i$$

where the exponential of the sum has been factored into the product of exponentials.

Next we expand the product as:

$$\prod_{i<j}(1+f_{ij}) = 1 + \sum_{i<j}f_{ij} + \sum_{i<j}\sum_{k<l}f_{ij}f_{kl} + \cdots$$

The contributions to the second term are significant whenever pairs of particles are close together. Diagrammatically we may represent the contributions to the second term as:

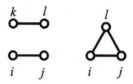

Contributions to the third term are significant either, if i, j, k, l are distinct, when pairs $i - j$ and $k - l$ are simultaneously close together or, if $j = k$ in the sums, when triples i, j, l are close together. The contributions to the third term may be represented as:

The contributions to the higher order terms may be represented in a similar way. The general expansion in this way is called a "cluster expansion" for obvious reasons.

3.1.3. *Low density approximation*

In the case of a dilute gas, we only need to consider the effect of pairwise interactions — the first two terms. This is because the probability of two given pairs being simultaneously close is small and the probability of three atoms being close is also small. Then we have

$$\prod_{i<j}(1+f_{ij}) \approx 1 + \sum_{i<j}f_{ij}$$

so that, within this approximation,

$$Q_N = \frac{1}{V^N}\int\left\{1 + \sum_{i<j}f_{ij}\right\}\mathrm{d}^{3N}q_i$$

$$= 1 + \int\sum_{i<j}f_{ij}\,\mathrm{d}^{3N}q_i.$$

There are $N(N-1)/2$ terms in the sum since we take all pairs without regard to order. Since the particles are identical, each integral in the sum will be the same, so that

$$Q = 1 + \frac{N(N-1)}{2V} \int f_{12} d^3 r_{12}.$$

The V^N in the denominator has now become V since the integration over $i, j \neq 1, 2$ gives a factor V^{N-1} in the numerator.

Finally, then, we have the partition function for the interacting gas:

$$Z = Z_{id} \left\{ 1 + \frac{N(N-1)}{2V} \int \left[e^{-U(r)/kT} - 1 \right] d^3 r \right\}$$

and on taking the logarithm, the free energy is the sum of the non-interacting gas free energy and the new term

$$F = F_{id} - kT \ln \left\{ 1 + \frac{N(N-1)}{2V} \int \left[e^{-U(r)/kT} - 1 \right] d^3 r \right\}.$$

In this low density approximation the second term in the logarithm, which accounts for pairwise intractions, is much less than the first term — Otherwise the third and higher order terms would also be important. But if the second term is small then the logarithm can be expanded. Also, obviously $(N-1)$ can be approximated by N. Thus we obtain

$$F = F_{id} - kT \frac{N^2}{2V} \int \left[e^{-U(r)/kT} - 1 \right] d^3 r.$$

A more rigorous treatment of the cluster expansion technique, including the incorporation of the higher order terms, is presented in the article by Mullin.[1]

3.1.4. *Equation of state*

The pressure is found by differentiating the free energy:

$$p = - \frac{\partial F}{\partial V} \bigg|_{T,N}$$

$$= \frac{NkT}{V} - \frac{N^2 kT}{2V^2} \int \left[e^{-U(r)/kT} - 1 \right] d^3 r.$$

We see that the effect of the interaction $U(r)$ can be regarded as modifying the pressure from the ideal gas value. The net effect can be either attractive or repulsive; decreasing or increasing the pressure. This will be considered, for various model interaction potentials $U(r)$. However before that we consider a systematic way of generalising the gas equation of state.

3.2. The Virial Expansion

3.2.1. *Virial coefficients*

At low densities we know that the equation of state reduces to the ideal gas equation. A systematic procedure for generalising the equation of state would therefore be as a power series in the number density N/V. Thus we write

$$\frac{p}{kT} = \frac{N}{V} + B_2(T)\left(\frac{N}{V}\right)^2 + B_3(T)\left(\frac{N}{V}\right)^3 + \cdots.$$

The B factors are called *virial coefficients*. And B_n is called the nth virial coefficient. By inspecting the equation of state derived above, we see that this is equivalent to an expansion up to the second virial coefficient. We see that the second virial coefficient is given by

$$B_2(T) = -\frac{1}{2}\int \left[e^{-U(\mathbf{r})/kT} - 1\right]d^3r$$

Since $U(\mathbf{r})$ is spherically symmetric we can integrate over the angular coordinates, giving

$$B_2(T) = -2\pi\int_0^\infty r^2\left[e^{-U(r)/kT} - 1\right]dr$$

which should be "relatively" easy to evaluate once the form of the interparticle interaction $U(r)$ is known. It is also possible to evaluate higher order virial coefficients, but this becomes more difficult.

3.2.2. *Hard core potential*

(The reader is referred to Reichl[2] for further details of the models in the next three sections.)

The hard core potential is specified by

$$U(r) = \infty \quad r < \sigma$$
$$= 0 \quad r > \sigma.$$

Here the single parameter σ is the hard core radius. This is actually modelling the particles as "billiard balls". There is no interaction when the

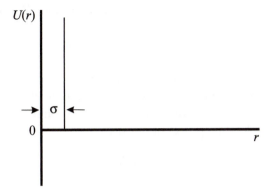

Fig. 3.1. Hard core potential.

particles are separated greater than σ and they are prevented, by the interaction, from getting closer than σ. It should, however, be noted that this model interaction is unphysical since it only considers the repulsive part; there is no attraction at any separation.

For this potential we have

$$e^{-U(r)/kT} = 0 \quad r < \sigma$$
$$= 1 \quad r > \sigma$$

so that the expression for $B_2(T)$ is

$$B_2(T) = 2\pi \int_0^\sigma r^2 dr$$
$$= \frac{2}{3}\pi\sigma^3.$$

In this case, we see that the second virial coefficient is independent of temperature, and it is always positive. The equation of state, in this case, is

$$pV = NkT\left\{1 + \frac{2}{3}\pi\sigma^3\frac{N}{V}\right\}$$

which indicates that the effect of the hard core is to increase the pV product over the ideal gas value.

It is instructive to rearrange this equation of state. Writing it as

$$pV \left\{ 1 + \frac{2}{3}\pi\sigma^3 \frac{N}{V} \right\}^{-1} = NkT,$$

we note that the correction term $\frac{2}{3}\pi\sigma^3 N/V$ is small within the validity of the derivation; it is essentially the hard core volume of a particle divided by the total volume per particle. So performing a binomial expansion we find to the same leading power of density

$$pV \left\{ 1 - \frac{2}{3}\pi\sigma^3 \frac{N}{V} \right\} = NkT$$

or

$$p \left\{ V - \frac{2}{3}N\pi\sigma^3 \right\} = NkT.$$

In this form we see that the effect of the hard core can be interpreted as simply reducing the available volume of the system. The "excluded volume" is *one half* the hard core volume of the particles. This is related to the fact that the $U(r)$ interaction is between *pairs* of particles and when summing over particles we must divide by two to avoid double counting of the energies.

3.2.3. *Square-well potential*

The square-well potential is made somewhat more realistic than the simple hard core potential by including a region of attraction as well as the

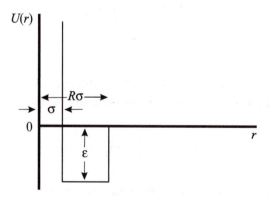

Fig. 3.2.   Square-well potential.

repulsive hard core. The potential is specified by

$$U(r) = \infty \qquad 0 < r < \sigma$$
$$= -\varepsilon \qquad \sigma < r < R\sigma$$
$$= 0 \qquad R\sigma < r$$

so we see that it depends on three parameters: σ, ε and R.

For this potential we have

$$e^{-U(r)/kT} = 0 \qquad 0 < r < \sigma$$
$$= e^{\varepsilon/kT} \qquad \sigma < r < R\sigma$$
$$= 1 \qquad R\sigma < \sigma$$

so that the expression for $B_2(T)$ is

$$B_2(T) = -2\pi \left\{ (-1) \int_0^\sigma r^2 dr + \left(e^{\varepsilon/kT} - 1 \right) \int_\sigma^{R\sigma} r^2 dr \right\}$$
$$= \frac{2}{3}\pi\sigma^3 \left\{ 1 - \left(R^3 - 1 \right) \left(e^{\varepsilon/kT} - 1 \right) \right\}.$$

In this case, using the more realistic potential, we see that the second virial coefficient depends on temperature. At low temperatures, where $B_2(T)$ is negative, this indicates that the attractive part of the potential is dominant and the pressure is reduced compared with the ideal gas case. And at higher temperatures, where it is intuitive that the small attractive part of the potential will be negligible, $B_2(T)$ will be positive and the pressure will be increased, as in the hard sphere case. The temperature at which $B_2(T)$ goes through zero is called the *Boyle temperature*, denoted by T_B. At very high temperatures we see from the expression for $B_2(T)$ that it will saturate at the value $2\pi\sigma^3/3$. Thus the general form of the second virial coefficient is as shown in Fig. 3.3.

3.2.4. *Lennard–Jones potential*

The Lennard–Jones potential is a very realistic representation of the inter-atomic interaction. It comprises an attractive $1/r^6$ term with a repulsive $1/r^{12}$ term. The form of the attractive part is well-justified as a description of the attraction arising from fluctuating electric dipole moments. The repulsive term is simply a power law approximation to the effect of the

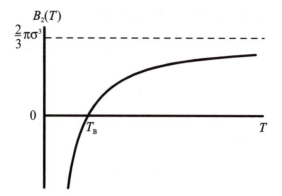

Fig. 3.3.   Second virial coefficient as a function of temperature.

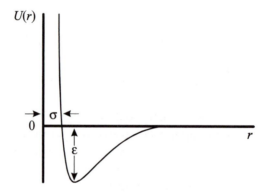

Fig. 3.4.   Lennard–Jones potential.

overlap of electronic orbits. We write the Lennard–Jones potential as

$$U(r) = 4\varepsilon \left\{ \left(\frac{\sigma}{r} \right)^{12} - \left(\frac{\sigma}{r} \right)^{6} \right\}$$

which depends on the two parameters: ε and R.

The integral for the second virial coefficient is

$$B_2(T) = -2\pi \int_0^\infty r^2 \left[e^{-\frac{4\varepsilon}{kT}\left\{\left(\frac{\sigma}{r}\right)^{12} - \left(\frac{\sigma}{r}\right)^6\right\}} - 1 \right] dr$$

which cannot be evaluated analytically. However, we can simplify somewhat by substituting $x = r/\sigma$ and $T^* = kT/\varepsilon$ and integrating by parts.

In this way we find

$$B_2(T) = \frac{2}{3}\pi\sigma^3 \frac{4}{T^*} \int_0^\infty x^2 \left(\frac{12}{x^{12}} - \frac{6}{x^6} \right) e^{-\frac{4}{T^*}\left(\frac{1}{x^{12}} - \frac{1}{x^6} \right)} \mathrm{d}x.$$

This is a function of temperature. The most direct way to proceed from here is to expand the exponential and integrate term by term. This gives a series for the second virial coefficient in inverse powers of temperature:

$$B_2(T) = -2 \left(\frac{2}{3}\pi\sigma^3 \right) \sum_{n=0}^\infty \frac{1}{4n!} \Gamma\left(\frac{2n-1}{4} \right) \left(\frac{1}{T^*} \right)^{(2n+1)/4}$$

where Γ is Euler's gamma function (the generalisation of the factorial function to non-integral argument).

Figure 3.5 shows the second virial coefficient as calculated from the Lennard–Jones potential and the square-well potential. It also shows experimental data for helium-4 and argon. Note that the calculated square-well curve saturates at high temperature at a value related to the hard core dimension. The data for helium indicate a slight reduction at higher temperatures, where the energetic collisions can cause the atoms to come even closer together. This is also shown for the Lennard–Jones calculated form; in that model, the hard core is not so hard. The deviations in the helium data at low temperatures are due to quantum effects.

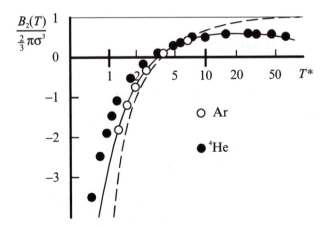

Fig. 3.5. Second virial coefficients: theory and experiment. The solid line is for the Lennard–Jones potential and the dashed line for square-well potential.

3.2.5. *Second virial coefficient for Bose and Fermi gas*

At low temperatures the equation of state of a quantum gas will depart from the ideal gas law. We saw this in Sec. 2.4.5. It is often stated that requirements of quantum statistics leads to "exchange forces" on classical "Boltzmann" particles; the exclusion principle for fermions gives a repulsive force while bosons experience an attraction. Such a view can, however, be seriously misleading.[3]

Since there is no real interaction potential, the method of the above calculations for the second virial coefficient is not appropriate. It is more sensible to go directly for the equation of state, approximated by expanding in powers of density. This is essentially what was done in Sec. 2.4.5 in considering the high temperature/low density limit of the quantum gas. There we obtained corrections to the ideal gas equation of state in powers of (the hypothetical) Fermi energy over kT. Upon substituting for the "Fermi energy" we find

$$pV = NkT \left\{ 1 \pm \frac{\pi^{3/2}}{2\alpha} \frac{N}{V} \frac{\hbar^3}{(mT)^{3/2}} \right\}$$

where the $+$ is for fermions and the $-$ is for bosons. Here α is the spin degeneracy factor. Thus, we may conclude

$$B_2(T) = \pm \frac{\pi^{3/2}}{2\alpha} \frac{\hbar^3}{(mT)^{3/2}}.$$

Note that this is monotonic in T; furthermore it is not analytic in $1/T$.

3.3. Thermodynamics

3.3.1. *Throttling*

In a throttling process, a gas is forced through a flow impedance such as a porous plug. For a continuous process, in the steady state, the pressure will be constant (but different) on either side of the impedance. When this happens to a thermally isolated system, so that heat neither enters nor leaves the system, then it is referred to as a Joule–Kelvin or Joule–Thompson process. This is fundamentally an irreversible process, but the arguments of thermodynamics are applied to such a system simply by considering the equilibrium initial state and the equilibrium final state which are applied way before and way after the actual process. This throttling process may be modelled by the diagram below.

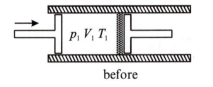 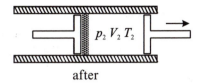

before   after

Fig. 3.6.   Joule–Kelvin throttling process.

Work must be done to force the gas through the plug. The work done is

$$\Delta W = -\int_{V_1}^{0} p_1 dV - \int_{0}^{V_2} p_2 dV = p_1 V_1 - p_2 V_2.$$

Since the system is thermally isolated, the change in the internal energy is due entirely to the work done:

$$E_2 - E_1 = p_1 V_1 - p_2 V_2$$

or

$$E_1 + p_1 V_1 = E_2 + p_2 V_2.$$

The *enthalpy H* is defined by

$$H = E + pV$$

thus we conclude that in a Joule–Kelvin process the enthalpy is conserved. The interest in the throttling process is that whereas for an ideal gas the temperature remains constant, it is possible to have either cooling or warming when the process happens to a non-ideal gas. The operation of most domestic refrigerators is based on this.

3.3.2.   *Joule–Thomson coefficient*

The fundamental differential relation for the enthalpy is

$$dH = TdS + Vdp.$$

It is, however, rather more convenient to use T and p as independent variables rather than the natural S and p. This is effected by expressing the entropy as a function of T and p whereupon its differential may be expressed as

$$dS = \frac{\partial S}{\partial T}\bigg|_{p} dT + \frac{\partial S}{\partial p}\bigg|_{T} dp.$$

But

$$\frac{\partial S}{\partial T}\bigg|_{p} = \frac{c_{p}}{T}$$

and using a Maxwell relation we have

$$\frac{\partial S}{\partial p}\bigg|_{T} = -\frac{\partial V}{\partial T}\bigg|_{p}$$

so that

$$dH = c_{p}dT + \left\{ V - T \frac{\partial V}{\partial T}\bigg|_{p} \right\} dp.$$

Now since H is conserved in the throttling process $dH = 0$ so that

$$dT = \frac{1}{c_{p}} \left\{ T \frac{\partial V}{\partial T}\bigg|_{p} - V \right\} dp$$

which tells us how the temperature change is determined by the pressure change. The *Joule–Thomson* coefficient μ_{J} is defined as the derivative

$$\mu_{J} = \frac{\partial T}{\partial p}\bigg|_{H},$$

giving

$$\mu_{J} = \frac{1}{c_{p}} \left\{ T \frac{\partial V}{\partial T}\bigg|_{p} - V \right\}$$

This is zero for the ideal gas (Problem 3.1). When μ_{J} is positive, then the temperature decreases in a throttling process when a gas is forced through a porous plug.

3.3.3.   *Connection with the second virial coefficient*

We consider the case where the second virial coefficient gives a good approximation to the equation of state. Thus we are assuming that the density is low enough so that the third and higher coefficients can be ignored.

This means that the second virial coefficient correction to the ideal gas equation is small and then solving for V in the limit of small $B_2(T)$ gives

$$V = \frac{NkT}{p} + NB_2(T)$$

so that the Joule–Thomson coefficient is then

$$\mu_J = \frac{NT}{c_p}\left\{\frac{dB_2(T)}{dT} - \frac{B_2(T)}{T}\right\}.$$

Within the low density approximation, it is appropriate to use the ideal gas thermal capacity

$$c_p = \frac{5}{2}Nk$$

so that

$$\mu_J = \frac{2T}{5k}\left\{\frac{dB_2(T)}{dT} - \frac{B_2(T)}{T}\right\}.$$

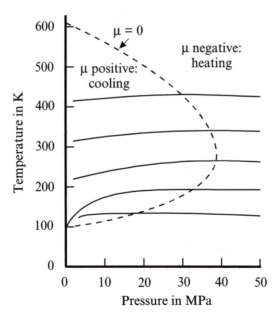

Fig. 3.7. Isenthalps and inversion curve for nitrogen.

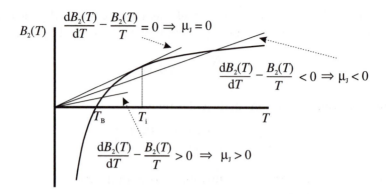

Fig. 3.8.   Behaviour of the Joule–Thomson coefficient.

3.3.4.   *Inversion temperature*

The behaviour of the Joule–Thomson coefficient can be seen from the following construction. We take the shape of $B_2(T)$ from the square-well potential model. While not qualitatively correct, this does exhibit the general features of a realistic interparticle potential.

We see that at low temperatures the slope of the curve, dB/dT is greater than B/T so that μ_J is positive, while at high temperatures the slope of the curve, dB/dT is less than B/T so that μ_J is negative. The temperature where μ_J changes sign is called the *inversion temperature*, T_i.

The inversion curve for nitrogen is shown as the dashed line in Fig. 3.7. We see that at high temperatures μ_J is negative, as expected. As the temperature is decreased, the inversion curve is crossed and μ_J becomes positive. Note, however that the low density approximation implicit in going only to the *second* virial coefficient keeps us away from the lower temperature region where the gas is close to condensing, where the Joule–Thomson coefficient changes sign again.

3.4.   Van der Waals Equation of State

3.4.1.   *Approximating the partition function*

Rather than perform an exact calculation in powers of a small parameter (the density), in this section we shall adopt a different approach by making an approximation to the partition function, which should be reasonably valid at *all* densities. Furthermore the approximation we shall develop will be based on the single-particle partition function. We shall, in this way, obtain an equation of state that approximates the behaviour of real gases. This equation was originally proposed by van der Waals in his Ph.D. Thesis in 1873. An English translation is available[4] and it is highly readable; van der Waals's brilliance shines out.

In the absence of an interaction potential the single-particle partition function is

$$z = \frac{V}{\Lambda^3}.$$

Recall that the factor V here arises from integration over the position coordinates. The question now is how to account for the interparticle interactions — in an approximate way. Now the interaction $U(r)$ comprises a repulsive hard core at short separations and an attractive long tail at large separations. And the key is to treat these two parts of the interaction in separate ways. The hard core excludes regions of space from the integration over position coordinates. This may be accounted for by replacing V by $V - V_{ex}$ where V_{ex} is the volume excluded by the hard core. The attractive long tail is accounted for by including a factor in the expression for z of the form

$$e^{-\langle E \rangle / kT}$$

where $\langle E \rangle$ is an average of the attractive part of the potential. Thus, we arrive at the approximation

$$z = \frac{V - V_{ex}}{\Lambda^3} e^{-\langle E \rangle / kT}.$$

Note that we have approximated the interaction by a *mean field* assumed to apply to individual particles. This allows us to keep the simplifying feature of the free-particle calculation where the many-particle partition function factorises into a product of single-particle partition functions. This is accordingly referred to as a mean field calculation.

3.4.2. *Van der Waals equation*

The equation of state is found by differentiating the free energy expression:

$$p = kT \left. \frac{\partial \ln Z}{\partial V} \right|_{T,N} = NkT \left. \frac{\partial \ln z}{\partial V} \right|_{T}.$$

Now the logarithm of z is

$$\ln z = \ln(V - V_{\text{ex}}) - 3\ln\Lambda - \langle E \rangle / kT$$

so that

$$p = NkT \left. \frac{\partial \ln z}{\partial V} \right|_T = \frac{NkT}{V - V_{\text{ex}}} - N\frac{\mathrm{d}\langle E \rangle}{\mathrm{d}V}$$

since we allow the average interaction energy to depend on volume (density). This equation may be rearranged as

$$p + N\frac{\mathrm{d}\langle E \rangle}{\mathrm{d}V} = \frac{NkT}{V - V_{\text{ex}}}$$

or

$$\left(p + N\frac{\mathrm{d}\langle E \rangle}{\mathrm{d}V} \right)(V - V_{\text{ex}}) = NkT.$$

This is similar to the ideal gas equation except that the pressure is increased and the volume decreased from the ideal gas values. These are constant parameters. They account, respectively, for the attractive long tail and the repulsive hard core in the interaction. Conventionally we express the parameters as aN^2/V^2 and Nb, so that the equation of state is

$$\left(p + a\frac{N^2}{V^2} \right)(V - Nb) = NkT$$

and this is known as the van der Waals equation.

Some isotherms of the van der Waals equation are plotted in Fig. 3.9 for three temperatures $T_1 > T_2 > T_3$. On the right-hand side of the plot,

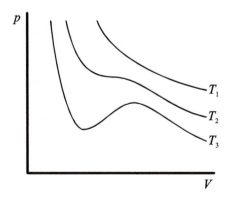

Fig. 3.9.   van der Waals isotherms.

corresponding to low density, we have gaseous behaviour; here the van der Waals equation gives small deviations from the ideal gas behaviour. On the left-hand side, particularly at lower temperatures, the steep slope indicates incompressibility. This is indicative of liquid behaviour. The non-monotonic behaviour at low temperature is peculiar and indeed it is non-physical. This will be discussed in detail in the next chapter.

The van der Waals equation gives a good description of the behaviour of both gases and liquids. In introducing this equation of state we said that the method should treat both low-density and high-density behaviour, and this has been done admirably. For this reason, Landau and Lifshitz[5] refer to the van der Waals equation as an interpolation equation. The great power of the equation, however, is that it also gives a good *qualitative* description of the gas–liquid phase transition, to be discussed in Chapter 4.

3.4.3. *Microscopic "derivation" of parameters*

In the mean field discussion the repulsive and attractive parts of the inter-particle interaction were treated separately. Within the spirit of this, let us consider how the two parameters of the van der Waals equation might be related to the two parameters of the Lennard–Jones inter-particle inter-action potential. The repulsion is strong; particles are correlated when they are very close together. We accounted for this by saying that there is zero probability of two particles being closer together than σ. Then, as in the hard core discussion of Sec. 3.2.1, the region of co-ordinate space is excluded, and the form of the potential in the excluded region ($U(r)$ very large) does not enter the discussion. Thus, just as in the discussion of the hard core model, the excluded volume will be

$$V_{\text{ex}} = \frac{2}{3}N\pi\sigma^3,$$

one half of the total hard core volume.

The attractive part of the potential is weak. Here, there is very little correlation between the positions of the particles; we therefore treat their distribution as approximately uniform. The mean interaction for a single pair of particles $\langle E_1 \rangle$ is then

$$\langle E_1 \rangle = \frac{1}{V}\int_\sigma^\infty 4\pi r^2 U(r)\mathrm{d}r$$

$$= \frac{1}{V}\int_\sigma^\infty 4\pi r^2 4\varepsilon \left\{\left(\frac{\sigma}{r}\right)^{12} - \left(\frac{\sigma}{r}\right)^6\right\}\mathrm{d}r = -\frac{8\pi\sigma^3}{3V}\varepsilon.$$

Now there are $N(N-1)/2$ pairs, each interacting through $U(r)$, so neglecting 1, the total mean energy per particle is

$$\langle E \rangle = \langle E_1 \rangle N/2$$

$$= -\frac{4\pi\sigma^3}{3}\frac{N}{V}\varepsilon.$$

In the van der Waals equation, it is the derivative of this quantity we require. Thus we find

$$N\frac{d\langle E \rangle}{dV} = \frac{4}{3}\pi\sigma^3 \left(\frac{N}{V}\right)^2 \varepsilon.$$

These results give the correct assumed N and V dependence of the parameters used in the previous section. So finally, we identify the van der Waals parameters a and b as

$$a = \frac{4}{3}\pi\sigma^3\varepsilon$$

$$b = \frac{2}{3}\pi\sigma^3.$$

3.4.4. *Virial expansion*

It is a straightforward matter to expand the van der Waals equation as a virial series. We express p/kT as

$$\frac{p}{kT} = \frac{N}{V - Nb} - \frac{aN^2}{kTV^2}$$

$$= \left(\frac{N}{V}\right)\left(1 - b\frac{N}{V}\right)^{-1} - \frac{a}{kT}\left(\frac{N}{V}\right)^2$$

and this may be expanded in powers of N/V to give

$$\frac{p}{kT} = \left(\frac{N}{V}\right) + \left(\frac{N}{V}\right)^2\left(b - \frac{a}{kT}\right) + \left(\frac{N}{V}\right)^3 b^2 + \left(\frac{N}{V}\right)^4 b^3 + \cdots.$$

Thus we immediately identify the second virial coefficient as

$$B_2^{\text{VW}}(T) = b - \frac{a}{kT}.$$

This has the form as sketched for the square-well potential. For this model we can find the Boyle temperature and the inversion temperature:

$$T_{\text{B}} = \frac{a}{bk},$$

$$T_{\text{i}} = \frac{2a}{bk}.$$

So we conclude that for the van der Waals gas the inversion temperature is double the Boyle temperature.

Incidentally, we observe that the third and all higher virial coefficients, within the van der Waals model, are constants independent of temperature.

3.5. Other Phenomenological Equations of State

3.5.1. *The Dieterici equation*

The Dieterici equation of state is one of a number of purely phenomenological equations crafted to give reasonable agreement with the behaviour of real gases. The Dieterici equation may be written as

$$p(V - Nb) = NkTe^{-\frac{Na}{kTV}}.$$

As with the van der Waals equation, this equation has two parameters, a and b, that parameterise the deviation from ideal gas behaviour. The interest in the Dieterici equation is that it gives a reasonable description of the behaviour of fluids in the vicinity of the critical point. This will be discussed in Chapter 4, in Sec. 4.2. For the present, we briefly examine the virial expansion of the Dieterici equation. In other words we will look at the way this equation treats the initial deviations from the ideal gas.

3.5.2. *Virial expansion*

In order to obtain the virial expansion we express the Dieterici equation as

$$\frac{p}{kT} = \frac{N}{V - Nb}e^{-\frac{Na}{kTV}}.$$

And from this, we may expand to give the series in N/V

$$\frac{p}{kT} = \frac{N}{V} + \left(\frac{N}{V}\right)^2\left(b - \frac{a}{kT}\right) + \left(\frac{N}{V}\right)^3\left(b^2 - \frac{a^2}{2k^2T^2} - \frac{ab}{kT}\right) + \cdots$$

This gives the second virial coefficient to be

$$B_2^D = b - \frac{a}{kT}.$$

This is the same as that for the van der Waals gas, and the parameters a and b may thus be identified with those of the van der Waals model. As a consequence, we conclude that both the van der Waals gas and the

Dieterici gas have the same values for the Boyle temperature and the inversion temperature.

The third virial coefficient is given by

$$B_3^D(T) = b^2 - \frac{a^2}{2k^2 T^2} - \frac{ab}{kT};$$

we see that this depends on temperature, unlike that for the van der Waals equation, which is temperature-independent.

3.5.3. *The Berthelot equation*

As with the Dieterici equation, the Berthelot equation is another of phenomenological origin. The equation is given by

$$\left(p + \frac{\alpha N^2}{kTV^2}\right)(V - Nb) = NkT.$$

The parameters of the Berthelot equation are given by α and b. We observe this equation is very similar to the van der Waals equation; there is a slight difference in the pressure-correction term that accounts for the long distance attraction of the intermolecular potential.

Since the Berthelot and van der Waals equation are related by $a = \alpha/kT$ it follows that the Berthelot second virial coefficient is given by

$$B_2^B = b - \frac{\alpha}{(kT)^2}.$$

Problems

3.1. Show that the Joule–Kelvin coefficient is zero for an ideal gas.

3.2. Derive the second virial coefficient expression for the Joule–Kelvin coefficient

$$\mu_J = \frac{2T}{5k}\left\{\frac{dB_2(T)}{dT} - \frac{B_2(T)}{T}\right\}$$

and find the inversion temperature for the square-well potential gas.

3.3. For the van der Waals gas, show that $T_B = \frac{a}{bk}$ and $T_i = \frac{2a}{bk}$.

3.4. Show that for a van der Waals fluid the critical parameters are given by $V_c = 3Nb$, $p_c = \frac{a}{27b^2}$, $kT_c = \frac{8a}{27b}$. Express these critical quantities in terms of the microscopic interaction (Lennard–Jones) parameters ε and σ.

3.5. Show that, for a van der Waals fluid, the combination of critical parameters $p_c V_c / N k T_c$ takes the universal value 3/8.

3.6. Show that for the Dieterici fluid, the critical parameters are given by $V_c = 2Nb$, $p_c = \frac{a}{4b^2e^2}$, $kT_c = \frac{a}{4b}$, and that the universal combination $p_c V_c / N k T_c$ has the value $2/e^2 = 0.271$.

3.7. Show that the partition function for an interacting gas may be expressed as

$$Z = Z_{id} \frac{1}{V^N} \int e^{-\sum\limits_{i<j} U(q_i, q_j))/kT} \, d^{3N}q$$

where Z_{id} is the partition function for a non-interacting gas. In terms of this expression explain why the partition function of a *hard sphere* gas might be approximated by

$$Z = Z_{id} \left(\frac{V - Nb}{V} \right)^N.$$

3.8. Show that the approximate partition function for the hard sphere gas in the previous question leads to the equation of state $p(V - Nb) = NkT$. This is sometimes called the Clausius equation of state. Give a physical interpretation of this equation.

Show that the first few virial coefficients are given by $B_2(T) = b$, $B_3(T) = b^2$, $B_4(T) = b^3$, etc. These virial coefficients are independent of temperature. Discuss whether this is a fundamental property of the hard sphere gas, or whether it is simply a consequence of the *approximated* partition function.

3.9. For a general interatomic interaction potential $U(r)$ we may define an effective hard core dimension d by $U(d) = kT$. What is the significance of this definition? Show that for the Lennard–Jones potential of Sec. 3.2.4, d is given by

$$d = \sigma \left\{ \frac{2}{1 + \sqrt{1 + kT/\varepsilon}} \right\}^{1/6}.$$

Plot this to demonstrate that d is a very slowly varying function of temperature.

If you have access to a symbolic mathematics system such as *Mathematica* or *Maple*, show that at low temperatures

$$d \sim \sigma \left(1 - \frac{1}{24} \frac{kT}{\varepsilon} + \frac{19}{1152} \left(\frac{kT}{\varepsilon} \right)^2 - \cdots \right).$$

So how high must the temperature be so that d differs appreciably from its zero temperature value?

How does d vary for the hard core and the square-well potentials of Secs. 3.2.2 and 3.2.3?

3.10. The one-dimensional analogue of the hard sphere gas is an assembly of rods constrained to move along a line (the Tonks model). For such a gas of N rods of length l confined to a line of length L, evaluate the configuration integral Q_N. Show that in the thermodynamic limit the equation of state is

$$f(L - Nl) = NkT$$

where f is the force, the one-dimensional analogue of pressure.

Comment on the similarities and the differences from the hard sphere equation of state mentioned in Problem 3.8 (Clausius equation) and the van der Waals equation of state.

3.11. Compare the square-well and the van der Waals expressions for the second virial coefficient. Show that they become equivalent when the range of the square-well potential tends to infinity while its depth tends to zero.

References

[1] W. J. Mullin, A new derivation of the virial expansion, *Amer. J. Phys.* **40** (1972) 1473.
[2] L. E. Reichl, *A Modern Course in Statistical Mechanics*, 2nd ed. (Wiley, 1998).
[3] W. J. Mullin and G. Blaylock, Quantum statistics: is there an effective fermion repulsion or boson attraction? *Amer. J. Phys.* **71** (2003) 1223.
[4] J. S. Rowlinson and J. D. van der Waals, *On the Continuity of the Gaseous and the Liquid States*, Studies in Statistical Mechanics XIV, ed. J. L. Lebowitz (North Holland, Amsterdam, 1988).
[5] L. D. Landau and E. M. Lifshitz, *Statistical Physics* (Pergamon Press, 1980).

CHAPTER FOUR

PHASE TRANSITIONS

4.1. Phenomenology

4.1.1. *Basic ideas*

In Chapter 3 we saw how interactions can affect the behaviour of systems, modifying properties from those of the corresponding ideal (non-interacting) system. Nevertheless a gas was still recognisable as a gas, albeit with somewhat altered properties; the interactions did not change the fundamental nature of the system. There is, however, another consequence of interactions when, by altering a thermodynamic variable such as temperature, pressure, etc. there can suddenly occur a dramatic change in the system's properties; there is a transition to a qualitatively different state. We refer to this as a phase transition.

Phase transitions present a challenge to statistical mechanics. At the transition point the system exhibits, by definition, singular behaviour. As one passes through the transition the system moves between analytically distinct parts of the phase diagram. But how can this be? The thermodynamic behaviour is embodied in the partition function. Thus the partition function must contain the details of any phase transition; it should exhibit the singular behaviour. But the partition function is merely a sum of Boltzmann factors — exponentials of the energy, which must therefore be analytic. There is a paradox: *does* the partition function contain the description of a phase transition or does it not? This question was a source of worry to physicists. And it was debated at the van der Waals Centenary Conference in November 1937. When a vote was taken it transpired that opinion was approximately equally divided between those who believed the partition function did contain details of a sharp transition and those who believed it did not!

The paradox was resolved by H. Kramers. He suggested that the singular behaviour only appears when the *thermodynamic limit* is taken — that is, when N and V go to infinity while the density N/V remains constant. This view was vindicated through the work of T. D. Lee and C. N. Yang published in 1952[1] (see also the book *Statistical Mechanics* by Huang[2]). Lee and Yang considered the fugacity $z = e^{\mu/KT}$ as a *complex* quantity, a function of the complex inverse temperature. There are zeros in the complex z plane. Lee and Yang found that as the thermodynamic limit is approached, one or more zeros of this function moves towards the real axis. It is this that causes a zero in the partition function, resulting in its singular behaviour.

Interactions are responsible for phase transitions. (The Bose–Einstein condensation is the only exception to this; there a phase transition occurs in the absence of interactions.) As a very rough "rule of thumb", often the temperature of the phase transition is approximately related to the interaction energy by $kT \approx E_{\text{int}}$. Thus the exchange interaction between electronic spins $-\hbar J \mathbf{S}_1 \cdot \mathbf{S}_2$ leads to a ferromagnetic transition at the Curie temperature $T_C \approx \hbar J/k$. However, in some cases it might not be immediately clear how to quantify the "interaction". In the BCS theory of superconductivity, it is interactions between lattice vibrations and the electrons that result in the superconducting transition. The transition temperature is given by $kT = 1.14\hbar\omega_D e^{-1/\rho(0)V}$ where ω_D is the Debye frequency characterising the phonons, $\rho(0)$ is the density of electron states at the Fermi surface and V is the volume. Here $\hbar\omega_D$ may be regarded as the characteristic energy, but the effect of the exponential is to shift the naively expected transition temperature by orders of magnitude; the rule of thumb is thus useless in this case.

From the reductionist perspective the first question one might ask about phase transitions is: "given the Hamiltonian for a system, can we predict whether there will be a phase transition?" The answer to this is: "probably in general no", for reasons that should become apparent. The second question might then be: "knowing that there is a transition and knowing the nature of that transition, can we predict the temperature of the transition from the Hamiltonian?" The answer to this question is that we can usually apply the rule above; often we can find a better approximation to the temperature of the transition; very occasionally we can find it exactly (analytically). The key point is the *nature* of the transition; that is difficult to predict.

Working against this reductionist view of phase transitions where the Hamiltonian is regarded as the all-important descriptor of the system, we have the observed phenomena of *universality*. It is found that many properties of systems in the vicinity of phase transitions do *not* depend on microscopic details, but are shared by dissimilar systems. We will find the explanation for this in the Landau picture of phase transitions where a common language for the phenomena is developed and in scaling arguments formalised in the renormalisation group.

4.1.2. *Phase diagrams*

Let us start with a very familiar example. The general behaviour of a *p–V* system is shown in Fig. 4.1. The variables of the thermodynamic configuration space are pressure *p* and volume *V* together with temperature *T*. A number of general observations can be made. There are regions where three distinct phases exist: solid, liquid and gas. And there are regions of two-phase coexistence.

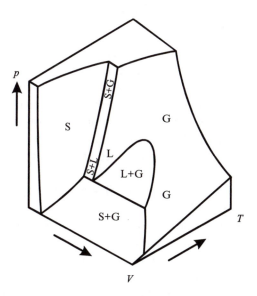

Fig. 4.1. Thermodynamic configuration diagram of a *p–V* system, *S* — solid, *L* — liquid, *G* — gas.

If we project the configuration diagram onto the p–T plane then the surfaces of coexistence become lines (why?). Such a phase diagram is shown in Fig. 4.2. Solid, liquid and gas regions are seen, separated by the coexistence lines. All three phases coexist at the *triple point* and the liquid–gas line terminates at the *critical point*. At the critical point, the distinction between the gas and the liquid phases disappears.

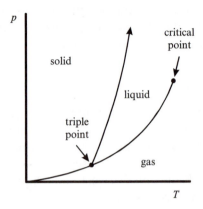

Fig. 4.2.   Phase diagram of a p–V system.

Many phase transitions involve a symmetry change. Often it is quite clear what symmetry is involved. Thus the liquid–solid and the gas–solid transitions both involve the breaking of translational symmetry. In the case of superfluids and superconductors it took many years until the appropriate symmetry was identified — gauge symmetry. However, in the case of the liquid–gas transition there is no symmetry broken. We will see what symmetries are broken in different types of phase transition in the next section.

The corresponding configuration diagram for a ferromagnetic magnetic system is shown in Fig. 4.3.

The phase diagram for the system is the projection of this diagram onto the B–T plane. This is shown in Fig. 4.4.

When the magnetic field is positive the magnetisation is pointing up. When the magnetic field is negative the magnetisation is pointing down. Above the critical temperature there is no magnetisation when B is zero, so one moves smoothly from the up state to the down state. But below the critical temperature there is a magnetisation in the absence of a magnetic

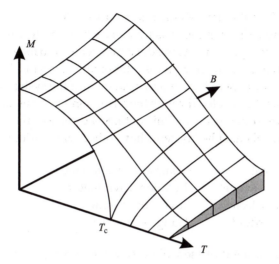

Fig. 4.3. Thermodynamic configuration diagram of a magnetic system.

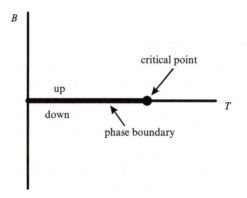

Fig. 4.4. Phase diagram for a magnetic system.

field. Then as B decreases through zero the magnetisation flips from up to down.

4.1.3. *Symmetry*

At the highest temperatures a system will be at its most disordered. It will have the highest symmetry possible, and this will be the symmetry of the system's Hamiltonian. As the temperature is lowered, when there is a phase transition, that symmetry is frequently broken; the resultant system will have a lower symmetry. Consider a magnet in zero magnetic field. The Hamiltonian contains $S_1 \cdot S_2$ which is rotationally invariant. Above the transition temperature there is no magnetisation and the system has *rotational symmetry*. But when cooled through the ferromagnetic transition a magnetisation appears spontaneously. This defines a direction and the rotational symmetry is broken. Other systems have other symmetries broken:

- Crystal   Translational symmetry
- Ferromagnet   Rotational symmetry
- Ferroelectric   Inversion symmetry
- Superfluid   Gauge symmetry

In general the ordered phase will possess a symmetry *lower* (i.e. less symmetric) than that of the system Hamiltonian.

Note, however, that the liquid–gas transition does not involve a change of symmetry; not all transitions involve symmetry breaking.

4.1.4. *Order of phase transitions*

Historically, the first stage in discussing phase transitions from a general perspective was the introduction of the idea of the *order* of the transition; this was the first step in classifying the nature of the non-analytic behaviour at the transition. When two phases coexist they have a common temperature and a common pressure (magnetic field). Thus the phases will each have the same Gibbs free energy.

In Fig. 4.5 we show the variation of the Gibbs free energy G for two different phases of a system — say solid and fluid, where the curves intersect. The equilibrium state will correspond to the lower G, so we see that the phase transition occurs at the point where the Gibbs free energy is the same for both phases. The observed G will thus display a kink at the phase transition.

Traditionally phase transitions were characterised, by Paul Ehrenfest, on the basis of the nature of the kink in G. If the nth derivative of G with respect to T (keeping other intensive variables constant) is discontinuous then it was said that the transition was nth order. Now since $\partial G/\partial T|_p = -S$ we see that the discontinuity in $\partial G/\partial T$ is the change in entropy between the two phases. Thus at a first-order transition the

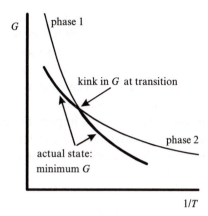

Fig. 4.5. Variation of Gibbs free energy for two phases.

entropy changes discontinuously, while at second- and higher order transitions the entropy varies continuously. And since the latent heat, the heat
absorbed, is given by $T\Delta S$ it follows that there *is* latent heat involved with
a first-order transition, but not with higher orders.

Nowadays the classification scheme is not used in quite this way;
there are only first- and second-order transitions. First-order transitions
are defined as above. And all transitions which are not first-order are
called second-order. So if there is latent heat involved in a phase transition
then it is first-order, otherwise it is second-order. Following the entropy
considerations above, first-order transitions are also called discontinuous
transitions, while so-called second-order transitions are also referred to as
continuous transitions. A different way of specifying the distinction will
be seen in the following section.

4.1.5. *The order parameter*

The modern general theory of phase transitions started with the work of
Landau in the 1930s. In developing a *general* treatment of phase transition
phenomena we must use a language applicable to all systems. On warming a ferromagnet (in zero field) the magnetisation goes to zero at the critical point. On warming a fluid along the coexistence curve the difference
between the liquid and gas densities goes to zero at the critical point. And
on warming a superfluid the superfluid density goes to zero at the critical
point. The general feature is that there is *some quantity* which goes to zero

at the critical point. When there is a symmetry broken on cooling through the critical point, this special quantity will be related to that symmetry. The special quantity is a measure of the order present in the system; it was called, by Landau, the *order parameter*. In the general case, we will use φ to denote the order parameter. For convenience, we will sometimes normalise φ to be unity in the fully-ordered state.

The nature of the order parameter is important; the magnetisation of a ferromagnet is a vector; the fluid density is a scalar. The order parameter for superfluid ^{4}He is a complex variable; those for liquid crystals and for superfluid ^{3}He are tensors.

System	Order parameter		
• Ferromagnet	magnetisation	$\mathbf{M}$	vector
• Ferroelectric	polarisation	$\mathbf{P}$	vector
• Fluid	density difference	$(n - n_c)$	real scalar
• Superfluid ^{4}He	ground state wavefunction	Ψ_0	complex scalar
• Superconductor	pair wavefunction	Ψ_s	complex scalar
• Ising	Ising 'magnetisation'	m	real scalar

We note that a complex scalar may, equivalently, be represented as a two-component vector.

The ferromagnet will be treated in Sec. 4.3 and again in Sec. 4.5.2. The ferroelectric is treated in Sec. 4.6. The fluid is covered in Sec. 4.2. Superfluids and superconductors are beyond the scope of this book. The Ising model is treated in Sec. 4.4. This is a magnet model where the interaction is restricted to the z direction. Its order parameter is a scalar. And there is a further magnet model where the interaction is restricted to the x–y plane. This is called the XY model; its order parameter is two-dimensional. It is mentioned briefly in Sec. 4.4.6.

The behaviour of the order parameter at the transition point gives a way to distinguish the order of the phase transition. As with the entropy, the order parameter changes discontinuously in a first-order transition, but continuously in a second-order transition; this reinforces the designation *discontinuous* and *continuous* transition. This behaviour is indicated in Fig. 4.6. Note that the apparent order of the transition may depend on the way the phase boundary is crossed. Thus while in general the liquid–gas transition is first-order, it is second-order when crossing the critical point from along the coexistence curve. Similarly, in the ferromagnet below T_c,

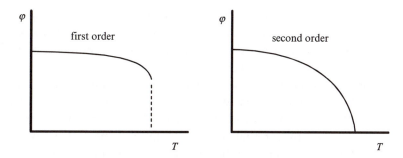

Fig. 4.6.   Order parameter in first- and second-order transitions.

if the field is varied through zero the magnetisation changes discontinuously, while it varies continuously when the field is zero and the transition is effected by varying the temperature. Furthermore, in the ferromagnet magnetisation can vary continuously from a positive to a negative value by going *around* the critical point — just as in a fluid.

Returning to the first question posed in Sec. 4.1.1 about predicting transitions from the system Hamiltonian, we stated in that section that the difficulty was in knowing the nature of the transition. This can now be reinterpreted as saying the difficulty is in identifying what the order parameter might be in the ordered phase.

Sometimes the order parameter will be one of the usual thermodynamic variables: magnetisation for the ferromagnet, polarisation for the ferroelectric and density for the fluid. However, it may be that the order parameter is some other — possibly unfamiliar — quantity such as the ground state wave function for superfluid ^{4}He or the pair wave function in superconductors.

A further important quantity is the thermodynamic variable *conjugate* to the order parameter. For the ferromagnet this is the magnetic field while for the ferroelectric it is the electric field. In the case of the fluid the conjugate variable is the chemical potential. The conjugate variable is important since by varying this quantity, we can couple to the order parameter in order to study the transition experimentally.

4.1.6.   *Conserved and non-conserved order parameters*

A new concept that we must introduce at this stage is that of conserved and non-conserved order parameters. In the magnetic transition, as we have seen, below the transition there is a spontaneously occurring magnetisation. This is the order parameter for that system. The value of the order parameter is determined from thermodynamic arguments by minimising the appropriate free energy.

In the case of the gas–liquid transition the order parameter is essentially the density (more precisely the difference between the density and its value at the critical point). Above the transition the density of the system is determined by external constraints: the volume and the number of atoms. Below the transition, the system separates into liquid and gas components. The liquid has a high density while the gas has a low density. However the *mean* density is still a fixed quantity; it is determined by the external constraints and not by minimising a free energy. The fluid system is thus said to have a *conserved order parameter*. We shall see that the binary mixture is another such system. The magnet, by contrast, has a *non-conserved* order parameter.

For a full treatment of systems with a conserved order parameter we must allow the order parameter to vary from place to place by introducing a position-dependent *order parameter density*.[3] It is then the integral of the order parameter density over the system's volume, which is the conserved quantity. The solution to this problem then gives the spatial dependence of the order parameter — in particular, it gives the variation at the interface between the different phases. There is a simpler approach, due to Gibbs,[4] that ignores the energy associated with the interface and simply considers coexisting regions with two different (uniform) values of the order parameter.

4.1.7. *Critical exponents*

In Sec. 4.1.1 we stated that many properties of systems in the vicinity of phase transitions do not depend on microscopic details, but are shared by dissimilar systems. In using a common description of phase transition phenomena, it is customary and convenient to specify the nature of the singular behaviour of quantities at the critical point. This is achieved through the introduction of *critical exponents* (and critical amplitudes). One of the successes of the modern theories of critical phenomena is in finding relations between the various critical exponents; this is achieved using scaling theory, as we shall see in the next section.

The first members of the critical exponent family are denoted by α, β, γ and δ. These describe the singularity of the heat capacity, order parameter, susceptibility and equation of state, respectively. In terms of the reduced temperature

$$t = \frac{T - T_c}{T_c}$$

they are defined (using the ferromagnet variables, for example) through

$$\text{heat capacity} \quad C \sim |t|^{-\alpha}$$
$$\text{order parameter} \quad m \sim |t|^{\beta}$$
$$\text{susceptibility} \quad \chi \sim |t|^{-\gamma}$$
$$\text{equation of state at} T_c \quad m \sim |B|^{1/\delta}.$$

Here the order parameter is the magnetisation density, the magnetic moment per unit volume.

There are two more critical exponents, which are connected with the spatial variation of fluctuations in the order as the critical point is approached. For this, we need to introduce the spatial correlation function for the order parameter. We define the correlation function

$$g(\mathbf{r}) = \langle m(\mathbf{r})m(0) \rangle$$

where $m(\mathbf{r})$ is the magnetisation density at position $\mathbf{r}$. The physical significance of correlation functions such as these is explained in Chapter 5 in Sec. 5.1.3.

In the vicinity of the critical point, the behaviour of the order parameter spatial correlation function is written as

$$g(\mathbf{r}) \sim r^{-p}e^{-r/l}.$$

And from this we obtain two more exponents, ν and η. These describe the divergence in the correlation length l and the power law decay p that remains at $t = 0$, when l has diverged. The exponents are defined through

$$\text{correlation length} \quad l \sim |t|^{-\nu}$$
$$\text{power law decay at } T_c \quad p = d - 2 + \eta$$

where d is the dimensionality of the system (it will turn out to be of interest to consider systems of dimensionality other than three).

A comprehensive theory of critical phenomena will give values for the critical exponents. However it turns out that there are fundamental relations between the critical exponents, so that of the six exponents only two are independent. This is understood from *scaling theory*, which we will encounter in the next section.

4.1.8. *Scaling theory*

Scaling theory involves the application of *dimensional analysis* to the study of the critical point. Near the critical point there are large fluctuations in the order parameter of the system. A dramatic example of this is the critical opalescence observed in fluids. This is a consequence of the density fluctuations that become very large at the critical point. As the critical point is approached, the fluctuations occur over longer and longer distances; this is the correlation length l referred to in the previous section.

Now l is an important length parameter in the system, and it becomes of macroscopic magnitude near the critical point. Since this length is macroscopic, it implies that microscopic details of the system become unimportant close to the critical point. *Thus* there is universality in critical phenomena.

Scaling theory relies on the hypothesis that close to the critical point the anomalous part of all quantities with the dimension of length will be proportional to the characteristic length l. And quantities of the dimension [length]^n will correspondingly be proportional to l^n. We use the notation [...] to denote the dimensions of a quantity. We consider the dimensionality some of the quantities that appear in, or lead to, the critical exponent definitions. A clear account of scaling theory is given in Huang's *Statistical Mechanics*.[2]

(a) Thermal capacity critical exponent α

Let us start with the Gibbs free energy G, which, by differentiating twice with respect to temperature, will give the thermal capacity. Now G/kT is dimensionless, although extensive, so that $g = G/kTV$, which is finite in the thermodynamic limit, has the dimensions of inverse volume:

$$[g] = L^{-d}$$

where d is the dimensionality of the system.

The scaling hypothesis is that L is proportional to the critical length l, whose critical exponent is defined to be ν:

$$l \sim |t|^{-\nu}$$

so that in the vicinity of the critical point we expect

$$g \sim |t|^{\nu d}.$$

Now thermal capacity is the second derivative of the Gibbs free energy with respect to temperature. So the definition of the exponent α:

$$C \sim |t|^{-\alpha}$$

implies the critical behaviour of g can also be written as

$$g \sim |t|^{2-\alpha}.$$

Equating these two forms for the exponent of g then gives

$$\alpha = 2 - \nu d,$$

showing how the heat capacity critical exponent relates to the critical length exponent and the system dimensionality.

(b) Order parameter critical exponent β

Now consider the order parameter — in the ferromagnet case this is the magnetisation per unit volume. The critical exponent β is defined through

$$m \sim |t|^{\beta}.$$

But we can also express the dimensionality of the order parameter from the spatial correlation function (correlation functions and their significance will be discussed in Chapter 5)

$$g(\mathbf{r}) = \langle m(\mathbf{r})m(0) \rangle \sim r^{-p} e^{-r/l}.$$

Thus

$$[m^2] = L^{-p}$$
$$= L^{-(d-2+\eta)}$$

so that

$$[m] = L^{(2-d-\eta)/2}.$$

Now according to the scaling hypothesis, that $L \sim |t|^{-\nu}$ then we have

$$m \sim |t|^{-\nu(2-d-\eta)/2}.$$

Equating these two forms for the exponent of m then gives

$$\beta = -\nu(2 - d - \eta)/2,$$

showing how the order parameter critical exponent relates to the critical length exponent, the power law decay exponent and the system dimensionality.

(c) Susceptibility critical exponent γ

Next we consider the susceptibility, which gives the critical exponent γ defined by

$$\chi \sim |t|^{-\gamma}.$$

Now the susceptibility is related to the order parameter's spatial correlation function $g(\mathbf{r})$ defined above. This may be seen from the fluctuation expression for the susceptibility obtained at the end of Sec. 2.7.2, Eq. (2.81):

$$\chi = \mu_0 \frac{1}{VkT} \langle M^2 \rangle.$$

The magnetisation M is given by the integral over the magnetisation density m. Thus in n dimensions

$$M = \int m(\mathbf{r}) \mathrm{d}^n r$$

and then

$$\langle M^2 \rangle = \int \mathrm{d}^n r_1 \int \mathrm{d}^n r_2 \langle m(\mathbf{r}_1) m(\mathbf{r}_2) \rangle.$$

On the assumption of translational invariance we can arbitrarily fix $\mathbf{r}_1$ to be at the origin and remove the first integral by simply multiplying by V, the n-dimensional volume

$$\langle M^2 \rangle = V \int \langle m(0) m(\mathbf{r}) \rangle \mathrm{d}^n r.$$

And then the susceptibility is given by

$$\chi = \mu_0 \frac{1}{kT} \int \langle m(0) m(\mathbf{r}) \rangle \mathrm{d}^n r.$$

This is a most important result. Our "derivation" relies on results of Chapter 2 for non-interacting particles. But the result is more general than that; thus our treatment should be regarded as no more than a plausibility argument.

The critical behaviour of the correlation function is given by

$$\langle m(\mathbf{r}) m(0) \rangle \sim r^{-p} e^{-r/l}.$$

then it follows that

$$[kT\chi] = L^{-p} \times L^d$$
$$= L^{2-\eta}$$

from the definition of p: $p = d - 2 + \eta$. So using the scaling hypothesis $L \sim |t|^{-\nu}$ we then have

$$kT\chi \sim |t|^{\nu(2-\eta)}.$$

Equating these two forms for the exponent of χ then gives

$$\gamma = \nu(2 - \eta),$$

showing how the susceptibility critical exponent relates to the critical length exponent and the power law decay exponent.

(d) Equation of state critical exponent δ

Finally, we look at the equation of state. From the definitions of β and δ we can write

$$m \sim |t|^{-\beta} \quad \text{and} \quad m \sim B^{1/\delta}$$

so then we have

$$B \sim |t|^{\beta\delta}.$$

Now B is related to the Gibbs free energy through

$$m = -\frac{\partial g}{\partial B}$$

so that

$$[B] = [g]/[m].$$

We have already found that $[g] = L^{-d}$ and $[m] = L^{(2-d-\eta)/2}$ so that

$$[B] = [g]/[m] = L^{-(2+d-\eta)/2}.$$

Using the scaling hypothesis $L \sim |t|^{-\nu}$ we then have

$$B \sim |t|^{\nu(2+d-\eta)/2}.$$

Equating these two forms for the exponent of B then gives

$$\delta = \nu(2 + d - \eta)/2\beta,$$

for the equation of state critical exponent.

158   *Topics in Statistical Mechanics*

By taking linear combinations of the above results we express the relations in the traditional manner as

$$\gamma = \nu(2 - \eta) \quad \text{Fisher law}$$
$$\alpha + 2\beta + \gamma = 2 \qquad \text{Rushbrooke law}$$
$$\gamma = \beta(\delta - 1) \quad \text{Widom law}$$
$$\nu d = 2 - \alpha \qquad \text{Josephson law}$$

The experimental verification of these results is strong evidence in favour of the scaling hypothesis. Thus it would appear that the correlation length *is* the only length of importance in the vicinity of the critical point. A consequence of the hypothesis is that only two critical exponents need be calculated for a specific system. Note the Josephson law is the only one to make explicit mention of the spatial dimensionality d.

4.1.9. *Scaling of the free energy*

The scaling hypothesis has important consequences for the free energy of systems in the vicinity of the critical point. In particular, it has important consequences for the *singular* part of the free energy. Following Fisher[5] we shall define the reduced free energy $f(T, B)$ as

$$f = \frac{-\Delta F}{kTV}$$

where ΔF is the deviation of the singular part of the free energy from its value at the critical point. We have divided by kT to make the quotient dimensionless. However it is still extensive, so we divide by the volume to produce an intensive quantity, with dimensions of inverse volume.

The scaling hypothesis is equivalent to the assumption that in the vicinity of the critical point the reduced free energy has the functional structure:

$$f(T, B) = A|t|^{2-\alpha} Y\left(D\frac{B}{|t|^{\Delta}}\right).$$

Here A and D are non-universal parameters that depend on the particular system. Essentially A sets the energy scale and D sets the magnetic field scale. The quantities α and Δ are the two universal exponents — recall the discussion of the previous section that there are only two independent exponents. Here α is the heat capacity exponent and Δ is related to the familiar critical exponents through $\Delta = 2 - \alpha - \beta$. The universal function

$Y(y)$ is defined so that $Y(0) = 1$; thus the need for the A prefactor. There are two branches to Y, one for $t > 0$ and one for $t < 0$. And as $y \rightarrow \infty$ the two branches must meet as t goes through zero.

From the above free energy function, all the scaling laws of the previous section may be derived[5]; see Problems 4.9 and 4.10. Thus we may regard the assumption that the free energy has the above form to be equivalent to the scaling hypothesis of the previous section.

4.2. First-Order Transition — An Example

In this section we shall consider the liquid–gas transition in a fluid. This is an example of a first-order transition with a conserved order parameter. The order parameter for this system is the density. The system has a conserved order parameter since its mean density is fixed (the volume is regarded as fixed).

4.2.1. *Coexistence*

In transitions involving a conserved order parameter the ordered phase can only evolve through spatial variation of the order parameter density — so that its integral over all space remains constant. Now a full solution of such a system would determine this spatial variation. However there is an approximation due to Gibbs[4] that sidesteps this complexity. The approximation must be justified *a posteriori*, but its basic assumption is that the system evolves into regions with two distinct values for the order parameter: two coexisting phases. At the boundary between the phases the order parameter will vary between that of the two distinct phases. The change will not be abrupt as that is energetically costly. Thus the Gibbs approach is a "thin wall" approximation where the fraction of particles in the intermediate regions is assumed to be extremely small.

When two phases coexist, then adding a small quantity of heat energy will result in the conversion of a small amount of the ordered phase to the disordered phase; there will be latent heat involved. Similarly, changing the volume at constant temperature will alter the proportions of the two phases. The coexistence region for a liquid–gas system is shown in Fig. 4.7. Clearly, during coexistence the pressure remains constant; the pressure depends only on the temperature.

A system will separate into two phases if it is energetically favourable to do so. Let us consider a fluid system held at constant temperature and

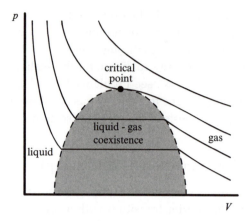

Fig. 4.7.   Coexistence region, showing p–V isotherms for liquid–gas system.

volume. Then the appropriate thermodynamic potential is the Helmholtz free energy $F(T, V)$. This free energy will be minimised when a constraint on the system is removed. In this case the "constraint" is the requirement that the system be uniform or homogeneous. We now remove this constraint and ask whether the free energy could be reduced through the system becoming inhomogeneous.

Figure 4.8 shows the Helmholtz free energy of a system at constant temperature as a function of *volume per particle* or specific volume v. The

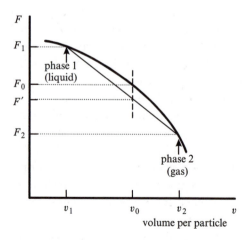

Fig. 4.8.   Helmholtz free energy curve.

solid curve in this figure shows the free energy that a homogeneous system would have. The system comprises N atoms or molecules occupying a volume V. The volume per particle of the homogeneous system is $v_0 = V/N$. We now ask if the system could lower its free energy by becoming inhomogeneous. In particular we explore the possibility that it segregates into regions of specific volume v_1 and v_2.

If this happens, it will be subject to the constraint that the total number of particles is fixed. Thus if a fraction α_1 of the particles is in regions of specific volume v_1 and a fraction $\alpha_2 = 1 - \alpha_1$ in regions of specific volume v_2 then

$$v_1 \alpha_1 + v_2 \alpha_2 = v_0$$

so that the fractions are given by

$$\alpha_1 = \frac{v_2 - v_0}{v_2 - v_1}, \quad \alpha_2 = \frac{v_0 - v_1}{v_2 - v_1}.$$

Then the free energy of the inhomogeneous system will be given by

$$\begin{aligned} F &= \alpha_1 F_1 + \alpha_2 F_2 \\ &= \frac{v_2 F_1 - v_1 F_2}{v_2 - v_1} - \frac{F_1 - F_2}{v_2 - v_1} v_0. \end{aligned}$$

This expression is linear in v_0. It is a linear interpolation between points (v_1, F_1) and (v_2, F_2). Thus the free energy of this inhomogeneous system will be given by the chord in Fig. 4.8, where it intersects the line $v = v_0$; this is indicated as F' in the figure.

We now have the condition for the system to remain homogeneous or to segregate. This depends upon whether F' is below F_0 or not; whenever F' falls below F_0 it will be favourable to separate into regions of different specific volume or density. Thus if the free energy curve is concave, as in Fig. 4.2, phase separation will occur, while if the curve is convex then it is favourable to remain homogeneous.

The lowest free energy is achieved when the straight line becomes a double tangent, as shown in Fig. 4.9. This determines the equilibrium state: having the lowest possible free energy for a given particle density. For obvious reasons it is referred to as the double tangent construction.

The two coexisting phases have a common tangent in the F–V plane. In other words $\partial F/\partial V|_T$ is the same for both phases. Now the differential

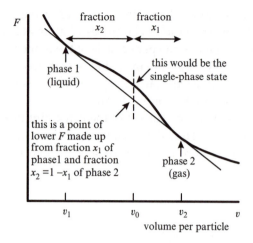

Fig. 4.9. Double tangent construction for phase coexistence.

relation for the Helmholtz free energy is

$$dF = -SdT - pdV$$

so that

$$p = -\left.\frac{\partial F}{\partial V}\right|_T;$$

the derivative is simply (minus) the pressure. Thus we conclude that in equilibrium the coexisting phases will have a common pressure — as expected.

4.2.2. *Van der Waals fluid*

We encountered the van der Waals equation of state in the previous chapter. We saw there how it provided a way of approximating the effects of inter-particle interactions in a mean-field manner. At low densities it provided a good description of gas-like behaviour while at high densities it provided a good description of liquid-like behaviour. We shall now see how this equation, when correctly interpreted, is also capable of providing a simple description for the gas–liquid transition. If we plot the equation of state

$$\left(p + \frac{aN^2}{V^2}\right)(V - Nb) = NkT$$

we obtain the following curves for different temperatures $T_1 < T_2 < T_3$.

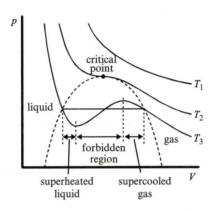

Fig. 4.10.   Isotherms of the van der Waals equation.

Below the critical point, the curve in the gas–liquid region appears strange. We know from the considerations of the previous section that it is preferable for phase separation to occur and the flat line is the usual behaviour. However, the curve represents the behaviour which would occur *if* the system remained homogeneous. There is a region of superheating and a region of supercooling. Although energetically unfavourable, it is possible in very clean systems that the new phase is not immediately nucleated. Then one can move down an isotherm from the liquid phase into the superheated region. Similarly one can move from the gas, up the isotherm, into the supercooled region. But when the curve changes direction you *have* to go to the two-phase state since the homogeneous phase is unstable if $\partial p/\partial V$ is positive. That would mean the pressure is increasing when the volume is increased!

4.2.3.   *The Maxwell construction*

The van der Waals isotherm is a curve, given by the van der Waals equation. However in the coexistence region the real isotherm must be a horizontal line; the pressure is a constant. Thus between points A and B, the

van der Waals curve is replaced by a straight line. The question is where
to position the line; what is the constant pressure during coexistence?

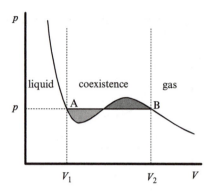

Fig. 4.11.   Maxwell construction on an isotherm.

Since $p = -\partial F/\partial V|_T$ it follows that the p–V curve of Fig. 4.11 is essentially
(minus) the derivative of the free energy graph of Fig. 4.9. In particular the
curved p–V isotherm comes from the derivative of the homogeneous system free energy curve, while the constant pressure coexistence isotherm
comes from the derivative of the double tangent line. From this the coexistence pressure may be determined in the following way.

The difference in the free energy between the pure liquid and the pure
gas, $F_1 - F_2$ may be found by integrating p with respect to V along the
isotherm:

$$F_1 - F_2 = \int_{V_2}^{V_1} p\,\mathrm{d}V.$$

Now we may integrate either along the straight line or along the curve
between points A and B of Fig. 4.11. If we integrate along the straight
line the free energy difference is the area under the straight line joining A
and B. And if we integrate along the curve, the free energy difference is
then the area under the van der Waals curve joining A and B. The free
energy must be the same either way and thus the area of the curve above
the straight line must be equal to the area below the line. This is known
as the *Maxwell construction*. It permits determination of the pressure of the
coexisting phases.

[When the two phases coexist they have equal temperature, pressure and chemical potential. We have accommodated temperature equality since we are talking about an isotherm. We have also accommodated pressure equality since this followed directly from the double tangent construction. It remains to consider the equality of chemical potentials of the two phases; this means that particles do not wish to flow from one phase to the other. We shall examine the consequence of the requirement that the chemical potential at point A be equal to that at point B, in Fig. 4.11. This will give a further insight into the Maxwell construction.

The chemical potential is the Gibbs free energy per particle, $\mu = G/N$. So let us find the change in chemical potential between points A and B and set this to zero. The differential expression for G is given by

$$dG = -SdT + Vdp.$$

So at constant temperature we then have

$$\Delta G = \int_{\text{point A}}^{\text{point B}} V dp.$$

It is convenient to reexpress this through integrating by parts:

$$\Delta G = pV\big|_{p,V_A}^{p,V_B} - \int_{V_A}^{V_B} p \, dV$$

$$= p(V_A - V_B) - \int_{V_A}^{V_B} p \, dV.$$

The first term is the area below the straight line joining A and B. The second term is the area below the van der Waals curve joining A and B. Thus ΔG is the difference between these two areas. And so the requirement that the chemical potentials are equal at points A and B again leads to the requirement that the area of the shaded part of the curve must be zero; we recover the Maxwell construction.]

The book by Mazenko[6] treats the Maxwell construction for the van der Waals fluid quite extensively.

4.2.4. *The critical point*

At the critical point, the distinction between the liquid and the gas disappears; the two phases become equivalent. This means that at the critical point the discontinuity in $\partial G/\partial T$ vanishes and thus the transition becomes second-order. In other words the first-order transition becomes second-order at the critical point. For the liquid–gas case if the density is fixed to be equal to that at the critical point, then as the system is cooled it will pass through the critical point to a state of two-phase coexistence.

Professionals in the business of phase transitions usually restrict their attention to what is happening in the vicinity of the critical point — and to the universality of the behaviour that emerges there, and thus the concentration is on critical exponents, etc.

The critical point for the liquid–gas system is the point of inflection, where

$$\left.\frac{\partial p}{\partial V}\right|_T = 0 \quad \text{and} \quad \left.\frac{\partial^2 p}{\partial V^2}\right|_T = 0.$$

The volume, pressure and temperature at this point are found, in terms of the van der Waals parameters a and b, to be

$$V_c = 3Nb, \quad p_c = \frac{a}{27b^2}, \quad kT_c = \frac{8a}{27b}.$$

4.2.5. *Corresponding states*

If we define reduced dimensionless variables v, π and t by

$$v = \frac{V}{V_c}, \quad \pi = \frac{p}{p_c}, \quad t = \frac{T}{T_c}$$

then the van der Waals equation takes on the *universal* form:

$$\left(\pi + \frac{3}{v^2}\right)\left(v - \frac{1}{3}\right) = \frac{8t}{3}.$$

This is universal in that there are no system-specific quantities. In other words, when the critical volume, temperature and pressure of a system are known then in terms of the reduced variables all fluids should obey the same equation. This is known as the *Law of Corresponding States*. Furthermore, the quantity p_cV_c/NkT_c is predicted to have the universal value $3/8 = 0.375$ for all liquid–gas systems.

As a demonstration of Corresponding States, Fig. 4.12 shows liquid–gas coexistence data for a number of substances plotted in reduced form, originally by Guggenheim.[7] And the points do indeed fall reasonably well onto a universal curve. However this is not the curve predicted from the van der Waals equation, the dotted curve in the figure.

We conclude that the law of corresponding states *does* seem to be followed, but the van der Waals equation does not give a good description of the universal behaviour.

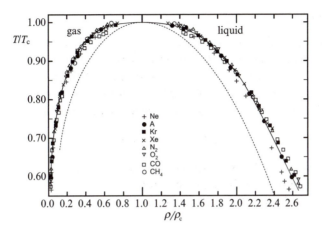

Fig. 4.12.   Liquid–gas coexistence.

The explanation of this is that the Law of Corresponding States is not reliant on the precise details of the van der Waals equation. It follows solely from the assumption that the energy of interaction between the particles is of the form

$$U(r) = \varepsilon f(r/\sigma)$$

where the energy parameter ε and the distance parameter σ are different for different substances, but the functional form of f is the same. In other words, the assumption is that the interaction is of a universal form with a scaling *energy* and a scaling *length* characterising the substance. So everything which follows from this interaction, such as the equation of state, the critical quantities, etc. must be functions only of these two parameters. Thus the universality when working in terms of the reduced variables, even if this is not precisely in accordance with the predictions of the van der Waals equation. Furthermore, since p_c, V_c and T_c are simply functions of ε and σ we can eliminate the two variables from the three equations so that the dimensionless quantity $p_c V_c / NkT_c$ is independent of ε and σ. While the van der Waals equation gives 0.375 for this quantity, experimental measurements give a value of 0.292 ± 0.002 for many substances.

	^{4}He	Ne	A	Kr	Xe	N_2	O_2	CO	CH_4
$p_c V_c / NkT_c$	0.308	0.305	0.291	0.290	0.278	0.292	0.292	0.294	0.280

Values of $p_c V_c / NkT_c$ for various substances.

4.2.6. *Dieterici's equation*

We encountered the Dieterici equation in Sec. 3.5.1. This is a purely phenomenological equation with far less microscopic justification than the
van der Waals equation. However, as stated in Chapter 3, the Dieterici
equation gives a better description of behaviour near the critical point than
does the van der Waals equation. We write the equation in the form

$$p\left(V - Nb\right) = NkTe^{-\frac{Na}{kTV}}$$

where, as we saw, the parameters a and b have precisely the same interpretation as in the van der Waals equation.

The critical point for the liquid–gas system is the point of inflection,
where

$$\left.\frac{\partial p}{\partial V}\right|_T = 0 \quad \text{and} \quad \left.\frac{\partial^2 p}{\partial V^2}\right|_T = 0.$$

The volume, pressure and temperature at this point are found to be

$$V_c = 2Nb, \quad p_c = \frac{a}{4b^2}e^{-2}, \quad kT_c = \frac{a}{4b}.$$

Then in terms of the reduced dimensionless variables v, π and t

$$v = \frac{V}{V_c}, \quad \pi = \frac{p}{p_c}, \quad t = \frac{T}{T_c}$$

the Dieterici equation takes on the universal form:

$$\pi\left(v - \frac{1}{2}\right) = \frac{t}{2}e^2e^{-\frac{2}{tv}}.$$

As with the reduced van der Waals equation, this is universal in that there
are no system-specific quantities. In other words, when the critical volume, temperature and pressure of a system are known, then in terms of
the reduced variables all fluids should obey the same equation.

For the Dieterici equation the critical point ratio is given by

$$\frac{p_c V_c}{NkT_c} = 2e^{-2}$$

$$= 0.271.$$

This is closer to the experimental 0.292 than the van der Waals value of
0.375. It is for this reason that one says the Dieterici equation gives a better
description of a fluid's behaviour in the vicinity of the critical point.

4.2.7. *Quantum mechanical effects*

The observant reader will notice in Fig. 4.12 that the data points for *liquid* neon show a small but consistent deviation from the universal line. Neon is the lightest molecule shown in the figure and its mass is the key. This may be seen even more dramatically in the behaviour of helium, which is significantly lighter. In Fig. 4.13, we show the Guggenheim plot coexistence data again, this time augmented by points from the two helium isotopes.[8]

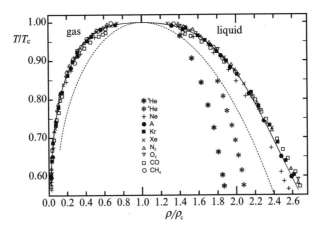

Fig. 4.13.   Liquid–gas coexistence including data for helium.

The breakdown of Corresponding States occurs as a consequence of quantum effects; this happens when the thermal deBroglie wavelength Λ becomes comparable with the interparticle separation. Then it is no longer appropriate to treat the particles as classical objects. We encountered Λ in Chapter 2 where we found

$$\Lambda = \hbar\sqrt{\frac{2\pi}{mkT}}.$$

Here m is the mass of the particles. This shows how Λ increases when m is small. At the critical point we can write Λ as

$$\Lambda_c = \frac{3}{4}\hbar\sqrt{\frac{6\pi}{m\varepsilon}}$$

where ε is the energy parameter of the Lennard–Jones potential, since

$$kT_c = \frac{16}{27}\varepsilon.$$

Then the ratio Λ_c/σ compares the thermal deBroglie wavelength with the interparticle spacing in the vicinity of the critical point; σ is the length parameter of the Lennard–Jones potential. We tabulate values of this ratio for most substances in Fig. 4.13.

	^{3}He	^{4}He	Ne	A	Kr	Xe	N$_2$	O$_2$
Λ_c/σ	1.595	1.388	0.307	0.096	0.053	0.033	0.119	0.103

Values of the quantum parameter Λ_c/σ for various substances.

For most substances we see that the ratio is small; the thermal deBroglie wavelength is significantly smaller than the interparticle spacing and the particles therefore behave classically. Neon is marginal; here the thermal deBroglie wavelength is beginning to become significant. And the classical picture is entirely inappropriate in the case of helium; there quantum effects become crucial. We should also note that a full quantum treatment would include the effect of statistics; recall that ^{4}He obeys Bose–Einstein statistics whereas ^{3}He obeys Fermi–Dirac statistics. There is a quantum extension of the law of Corresponding States, introduced by de Boer[9] in 1948.

4.3.   Second-Order Transition — An Example

4.3.1.   *The ferromagnet*

The essential phenomenon associated with the ferromagnet is that below a certain temperature a magnetisation will spontaneously appear in the absence of an applied magnetic field. It is understood that the interaction responsible for ferromagnetism is the exchange interaction between electron spins. The origin of the exchange interaction is the necessity to antisymmetrise the electronic wavefunction, together with the Coulomb repulsion between electrons. Then the symmetric and the antisymmetric wavefunctions have different energies and this may be written as an *effective* spin-dependent Hamiltonian:

$$\mathcal{H}_x = -\hbar J \sum_{i,j}^{\text{nn}} \mathbf{S}_i \cdot \mathbf{S}_j$$

called the *Heisenberg* Hamiltonian. You should be familiar with this from your Atomic Physics studies. The essential feature of this interaction is that when J is positive the energy is minimised when the spins are parallel; this is the energetically favourable state. When J is negative the favourable state occurs when neighbouring spins are antiparallel; this is an antiferromagnet. Note that the exchange interaction is rotationally invariant. The sum is over nearest neighbours and J is called the exchange frequency.

As written above, the sum counts each pair twice since both i and j vary freely. For this reason we will sometimes use the alternate way of expressing the same thing:

$$\mathcal{H}_x = -2\hbar J \sum_{i<j}^{\text{nn}} \mathbf{S}_i \cdot \mathbf{S}_j$$

since in this case the restriction $i < j$ avoids the double counting.

The phase diagram for a ferromagnet is shown in Fig. 4.14. The system is restricted to the surface, and there is a second, lower surface at negative B and M. The surface is symmetric under $M \to -M$, $B \to -B$. Note that there is a smooth variation between positive and negative magnetisation only at $B = 0$ and for $T > T_c$. Otherwise the change is discontinuous.

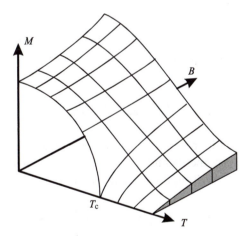

Fig. 4.14.   Phase diagram for a magnetic system.

For temperatures below the critical temperature a magnetisation isotherm is shown in Fig. 4.15. Since $B = 0$ when the magnetisation inverts, there is no cost in energy.

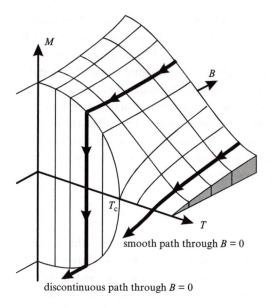

M

B

T_c

T

smooth path through *B* = 0

discontinuous path through *B* = 0

Fig. 4.15.   Variation of *M* as *B* goes through zero.

4.3.2.   *The Weiss model*

The behaviour of the ferromagnet is contained in its Hamiltonian. In the presence of a magnetic field **B**, assumed to point in the *z* direction the Hamiltonian is

$$\mathcal{H} = -\hbar\gamma\mathbf{B}\cdot\sum_i \mathbf{S}_i - \hbar J\sum_{i,j}^{nn}\mathbf{S}_i\cdot\mathbf{S}_j$$

where the first term is the usual $-\mathbf{M}\cdot\mathbf{B}$ term for the Zeeman interaction between magnetic moments and a magnetic field; γ is the magnetogyric ratio of the electron. Writing this Hamiltonian as

$$\mathcal{H} = -\hbar\gamma\left\{\mathbf{B} + \frac{J}{\gamma}\sum_j^{nn}\mathbf{S}_j\right\}\cdot\sum_i \mathbf{S}_i$$

we see that the effect of the exchange term is to contribute some extra magnetic field at the site of each spin due to its neighbours.

   The extra field at each site is different; it is a result of the particular disposition of the site's neighbours. Furthermore the field at a site will vary because of the time evolution of the neighbouring moments induced by the Hamiltonian. However the time-averaged field at different sites will be expected to be the same. In the model introduced by Pierre Weiss in 1907 the extra field is taken to be the same, and constant, throughout the specimen. The Weiss model is thus a mean field model. The mean extra magnetic field is then

$$\mathbf{b} = \frac{J}{\gamma} \left\langle \sum_j^{nn} \mathbf{S}_j \right\rangle$$
$$= nJ \langle \mathbf{S} \rangle / \gamma$$

where n is the number of nearest neighbours of a site.

We have, then, a single-particle effective Hamiltonian. The properties of this system may be found from the paramagnet model treated in Chapter 2. There we found the magnetisation (total magnetic moment) of an assembly of N spins of magnitude $\hbar/2$ is given by

$$\mathbf{M} = N\gamma\hbar \langle \mathbf{S} \rangle$$

so that

$$M = N\frac{\gamma\hbar}{2} \tanh \frac{\gamma\hbar B}{2kT}.$$

In the presence of a magnetic field the magnetisation $\mathbf{M}$ will be parallel to $\mathbf{B}$. Now we must add the mean extra field

$$\mathbf{b} = \frac{nJ}{\gamma} \langle \mathbf{S} \rangle = \frac{nJ}{N\gamma^2\hbar} \mathbf{M}$$

so that the magnetisation is then

$$M = N\frac{\gamma\hbar}{2} \tanh \frac{\gamma\hbar}{2kT} \left\{ B + \frac{nJ}{N\gamma^2\hbar} M \right\}.$$

This is an implicit equation relating magnetisation, magnetic field and temperature. It provides a mathematical representation of the phase diagram shown in the previous section. The equation is, however, non-linear and impossible to solve analytically. But we can find the spontaneous magnetisation and the critical temperature; we do this in the next section.

4.3.3. *Spontaneous magnetisation*

In zero applied magnetic field a magnetisation will spontaneously appear when the temperature falls below the critical temperature. This may be seen from the equation for magnetisation. When $B = 0$, this becomes

$$M = N\frac{\gamma\hbar}{2}\tanh\frac{nJ}{2kTN\gamma}M.$$

This is still a non-linear and implicit equation that is difficult to solve analytically. But we can adopt a graphical method, which is highly instructive. We define the auxiliary quantity X by

$$X = \frac{nJ}{2kTN\gamma}M$$

and then we have two simultaneous equations in two unknowns. The solution corresponds to the intersection of the curves representing the two equations:

$$M = N\frac{\gamma\hbar}{2}\tanh X$$

$$M = \frac{2kTN\gamma}{nJ}X.$$

These equations are plotted for three different temperatures in Fig. 4.16 below.

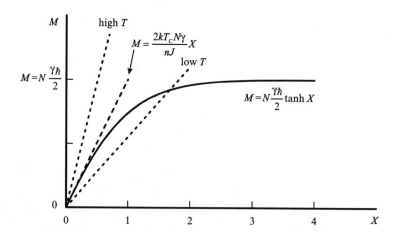

Fig. 4.16. Graphical solution for spontaneous magnetisation.

We observe there is always a solution at $M = 0$. At high temperatures, as expected, this is the only solution. At low temperatures, however, there is a second solution where the line intersects the curve at a non-zero value of M. We shall see below in Sec. 4.5.2 that this corresponds to the energetically favourable solution. Thus for temperatures below a critical value there will be a non-zero magnetisation. The critical temperature is that for which the straight line is tangential with the tanh at the origin, when

$$N\frac{\gamma\hbar}{2} = \frac{2kTN\gamma}{nJ},$$

that is,

$$T_c = \frac{\hbar n}{4k}J.$$

In the case of a 2d square lattice, for which $n = 4$, one then has

$$kT_c = \hbar J$$

an example of our 'rule of thumb' for the transition temperature mentioned in Sec. 4.1.1.

Although an analytic solution for M in terms of T is not possible, we can solve for T in terms of M. The magnetisation obeys the implicit equation

$$M = N\frac{\gamma\hbar}{2}\tanh\left\{\frac{nJ}{2kTN\gamma}M\right\}$$

which can be written as

$$M = M_0\tanh\left\{\frac{M}{M_0}\frac{T_c}{T}\right\}$$

where $M_0 = N\gamma\hbar/2$, the saturation magnetisation and $T_c = \hbar nJ/4k$, the critical temperature. The tanh equation may be inverted and solved in the following way. We write it as

$$\frac{M}{M_0}\frac{T_c}{T} = \tanh^{-1}\left(\frac{M}{M_0}\right)$$

$$= \frac{1}{2}\ln\left(\frac{1+M/M_0}{1-M/M_0}\right),$$

so that

$$\frac{T}{T_c} = \frac{2M/M_0}{\ln\left(\dfrac{1+M/M_0}{1-M/M_0}\right)}.$$

Admittedly, this gives the temperature in terms of the magnetisation rather than *vice versa*, but it is an explicit expression. When plotted, this gives the form for the variation of spontaneous magnetisation with temperature. This solution is shown in Fig. 4.17.

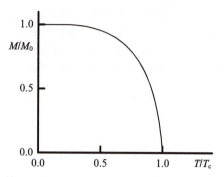

Fig. 4.17. Spontaneous magnetisation of a ferromagnet.

4.3.4.   *Critical behaviour*

In the vicinity of the critical point the magnetisation will be very small. In that case we may expand the tanh:

$$\frac{M}{M_0} = \frac{M}{M_0}\frac{T_c}{T} - \frac{1}{3}\left(\frac{M}{M_0}\frac{T_c}{T}\right)^3 + \cdots$$

(since $\tanh x = x - x^3/3 + \cdots$).

This expression may now be rearranged to give

$$\frac{M}{M_0} \sim \pm\sqrt{3}\frac{T}{T_c}\left(1 - \frac{T}{T_c}\right)^{1/2}$$

$$\sim \pm\sqrt{3}\left(1 - \frac{T}{T_c}\right)^{1/2}$$

which gives the behaviour of the magnetisation in the region of the critical point. This shows the initial growth of the magnetisation as the temperature is decreased through the critical point.

The dominant, singular, part of this behaviour is contained in the factor $(1 - T/T_c)^{1/2}$. In particular, it is the exponent $1/2$ which characterises how M "takes off" from zero. This is the critical exponent β introduced in Sec. 4.1.7. Thus our conclusion is that the Weiss mean field model gives a value of $1/2$ for β.

The coefficient of the dominant singular temperature behaviour, here the coefficient of $(1 - T/T_c)^{1/2}$, is called the *critical amplitude*. For the Weiss mean field model the magnetisation critical amplitude is thus seen to be $\sqrt{3}$.

4.3.5. *Magnetic susceptibility*

The equation of state for the non-interacting paramagnet is given by

$$M = N\frac{\gamma\hbar}{2}\tanh\frac{\gamma\hbar B}{2kT}$$

or, in terms of the saturation magnetisation $M_0 = N\gamma\hbar/2$:

$$M = M_0\tanh\frac{M_0 B}{NkT}.$$

At high temperatures/low polarisation the tanh is expanded to leading order, resulting in a linear relation between M and B

$$M = \frac{M_0^2 B}{NkT}.$$

In this case one can define a *magnetic susceptibility* χ

$$\chi = \frac{1}{\mu_0 V}\frac{M}{B}.$$

(The factor μ_0 is included to make χ dimensionless.) The susceptibility of the non-interacting paramagnet is then

$$\chi = \frac{M_0^2}{\mu_0 N^2 k}\frac{1}{T}.$$

The $1/T$ dependence of the susceptibility is known as Curie's law.

We now include the effect of the mean field, which can be written (in terms of T_c) as

$$b = \frac{Nk}{M_0^2}T_c M$$

so that now the magnetisation is

$$M = \frac{M_0^2}{NkT}\left(B + \frac{Nk}{M_0^2}T_c M\right)$$

$$= \frac{M_0^2}{NkT}B + M\frac{T_c}{T}.$$

This may be solved to give

$$M = \frac{M_0^2}{Nk(T - T_c)}B$$

so that the susceptibility is now

$$\chi = \frac{M_0^2}{\mu_0 V Nk(T - T_c)}.$$

This is similar to Curie's law except that the temperature T is replaced by $T - T_c$. By contrast to Curie's law this is referred to as the Curie–Weiss law.

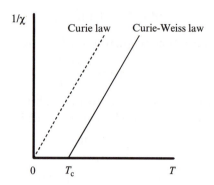

Fig. 4.18.   Curie and Curie–Weiss laws.

We see that by plotting the inverse susceptibility as a function of temperature, the transition temperature may be found by extrapolating to zero.

The Curie–Weiss susceptibility exhibits divergent behaviour as the critical point is approached. The exponent of $(T - T_c)^{-1}$ is 1. This is the critical exponent γ. Thus we conclude that the Weiss mean field model gives a value of 1 for the exponent γ.

4.3.6.   *Goldstone modes*

The essential feature of the ferromagnetic transition is that below a critical temperature a spontaneous magnetisation appears. This happens in the absence of an applied magnetic field. Thus the magnetisation appears in a completely arbitrary direction. The interaction responsible for the transition is the Heisenberg exchange Hamiltonian and this is rotationally invariant. Nevertheless, the vector magnetisation must point in some direction. And thus the transition breaks the symmetry of the Hamiltonian.

There is, of course, also the $M = 0$ solution and this does respect the symmetry of the Hamiltonian but the non-zero M solution, the symmetry-breaking solution, is energetically favourable. This means that the ground state of the system is highly degenerate. We may denote the set of degenerate states by $|\hat{\mathbf{r}}\rangle$, where $\hat{\mathbf{r}}$ is the unit vector pointing in the direction of the magnetisation. It is an observed fact that the ground state of this system is always one of the $|\hat{r}\rangle$ states; one never observes a linear superposition of such states, even though this is allowed by the laws of quantum mechanics. This paradox is discussed in Huang's Statistical Mechanics book,[2] p. 301. However this is essentially the classic paradox of Schrödinger's cat which goes to the very heart of quantum theory.

Fig. 4.19. Different possible ground states.

The ground state corresponds to a uniform order parameter; the magnetisation points in the same direction throughout the specimen. Now it costs energy to deform the order parameter. So if we consider a sinusoidal spatially varying order parameter $\mathbf{M}(\mathbf{r})$:

$$\mathbf{M}(\mathbf{r}) = M_0\{\cos(\mathbf{k}\cdot\mathbf{r})\hat{x} + \sin(\mathbf{k}\cdot\mathbf{r})\hat{y}\}$$

then this will have an energy above that of the ground state. This expression for $\mathbf{M}(\mathbf{r})$ is the spatial part of a spin wave with wavevector $\mathbf{k}$. In the limit $k \rightarrow 0$ the expression reduces to the uniform ground state with the magnetisation pointing in the x direction. Thus the energy of the spin wave goes to zero continuously as k goes to zero.

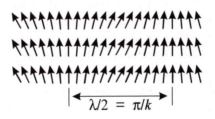

Fig. 4.20. Spatial variation of the order parameter.

This is the physical content of Goldstone's theorem which states that when a continuous symmetry is broken there will be excitations involving variations in the order parameter and the dispersion relation for these excitations, $\varepsilon(k)$, satisfies

$$\varepsilon(k) \to 0 \quad \text{as } k \to 0.$$

These excitations are known as Goldstone modes or, when quantised, as Goldstone bosons. They are the low temperature excitations and thus they determine the low temperature thermal properties of the system.

The fact that $\varepsilon(k)$ goes to zero continuously means that there is no energy gap between the ground state and the excitations. This may also be interpreted as saying that the Goldstone bosons have zero mass. And this property follows only when the interaction responsible for the symmetry breaking, here the exchange Hamiltonian, is of short range. We may contrast this with the superconducting transition. There the long range of the Coulomb force results in a gap in the plasmon excitation spectrum; then the bosons have mass. This is the condensed matter analogue of the Higgs mechanism of quantum field theory whereby elementary particles acquire mass.

Note that Goldstone modes only appear when a continuous symmetry is broken. The order parameter cannot vary continuously in space when it is a discrete symmetry that is broken, such as that for the Ising model treated in the next section.

4.4. The Ising and Other Models

4.4.1. *Ubiquity of the Ising model*

The Ising model is a simple Hamiltonian model introduced to treat ferromagnetism. It was originally proposed in the early 1920s by Wilhelm Lenz in an attempt to explain ferromagnetism from microscopic first principles. This was after Weiss had introduced the mean field model of ferromagnetism, but before the advent of the quantum Heisenberg exchange Hamiltonian (which we had used to motivate the Weiss mean field). This pre-quantum model comprises an assembly of magnetic moments each of which can point either parallel or antiparallel to a given direction. It was Lenz's hope that this very simplest of interacting systems would exhibit a ferromagnetic transition. Lenz set this as a problem to his student Ernst Ising. Ising was able to solve only the one-dimensional case and he was disappointed to discover that there was no phase transition. It was then not until the 1940s that the two-dimensional case was solved by Lars Onsager, giving a ferromagnetic transition at a finite temperature.

However, the three-dimensional Ising model is still not solved; probably it does not admit an analytic solution. A very readable account of the history of the Ising model is given in the review by Brush.[10]

Note that the (Weiss) mean field model and its predictions make no mention of the spatial dimension. Within that context the discovery of the difference between the one- and two-dimensional case of the Ising model was rather surprising.

The Ising model is one of the simplest descriptions of a system that leads to a phase transition. It is important furthermore because, notwithstanding its magnetic origin, it can be applied to a variety of dissimilar physical systems. This is an example of the phenomenon of *universality* in phase transitions. Perhaps of even more importance is the fact that the Ising model in two dimensions has been solved analytically. This is one of the very few microscopic models of interacting systems that have been solved exactly, to exhibit a phase transition.

The Ising model is specified in terms of a Hamiltonian, so it is a complete microscopic description — but it keeps this description as simple as possible. It is a lattice model; the moments are fixed on sites. Often interactions are permitted only between nearest neighbours. The order parameter is a scalar: the expectation value of the Ising "spin". The insight of Lenz was in stripping away all superfluous aspects of the ferromagnet while keeping the essence of the system that leads to a transition.

We will write the Ising Hamiltonian down in the next section. For the present, we simply note that this involves the energies of neighbouring pairs of the magnetic moments. We shall denote the directions in which the moments can point as up and down. Parallel neighbours have one energy, $\varepsilon_{\uparrow\uparrow}$, while antiparallel neighbours have another, $\varepsilon_{\uparrow\downarrow}$. In the case that

$$\varepsilon_{\uparrow\uparrow} < \varepsilon_{\uparrow\downarrow},$$

the parallel state is energetically favourable and the ordered phase will thus be ferromagnetic. But when

$$\varepsilon_{\uparrow\uparrow} > \varepsilon_{\uparrow\downarrow},$$

the antiparallel state is favoured and the ordered phase will be antiferromagnetic.

The Ising description need not, however, be restricted to real magnetic moments. For example, consider a binary mixture of A atoms and B atoms, as we shall do in Sec. 4.7. Here again the interaction energies depend upon neighbouring particles. We can imagine that an A atom is represented by an "up moment" and a B atom by a "down moment". The analogue of the ferromagnetic state is phase-separation, when the A atoms coalesce together as do the B atoms. However if it is favourable for unlike atoms to be neighbours, then we have the analogue of the antiferromagnet and the ordered phase will be a superlattice structure.

Finally, consider a fluid system modelled as a lattice gas. We interpret an up moment as a molecule on the site, while a down moment indicates an empty site: a hole. Adsorption on a surface is an example of this model in two dimensions.

We see how, by interpreting the parameters appropriately, the Ising model may be used to describe a variety of physical systems. In most of the following we will use the language of the magnetic case, but the generalisation to other systems is straightforward. However one should note the distinction between conserved and non-conserved order parameters. For the binary mixture the order parameter is conserved; the numbers of A and B atoms is fixed. However in the magnetic case the order parameter is not conserved as spins may flip. The lattice gas model has a fixed number of molecules. If the number of available sites is fixed then the order parameter is conserved. But if the number of available sites is unlimited then the order parameter is not conserved. In the magnetic description we may imagine applying a magnetic field to the system; this will determine the fraction of up and down moments. In the case of the non-conserved order parameter, this field is a free parameter. But for the conserved order parameter case this field is a "chemical potential" whose value is determined by the conserved quantity; it fixes the number of A and B atoms in the binary mixture or the number of molecules in the lattice gas model.

4.4.2. *Magnetic case of the Ising model*

In the magnetic Ising model we have a magnetic moment on each lattice site which may point either "up" or "down". As there are only two states, this is reminiscent of a spin $1/2$ problem. It is thus appropriate to utilise a spin operator description, taking the up direction as pointing along the z axis. We then assign to each site a spin with eigenvalue $S^z = +1/2$ if it carries an "up" moment and $S^z = -1/2$ if it carries a "down" moment.[*] We

---

[*]Traditional treatment of the Ising model ascribes the values $+1$ and -1 to the two values of the Ising "spin". We adopt a different convention that makes better connection with the spin $1/2$ quantum mechanical description of the Heisenberg magnet.

must count the number of parallel and antiparallel neighbour pairs. This may be effected through consideration of the product of two neighbouring spins' eigenvalues; it will have magnitude $+1/4$ if the neighbours are parallel and $-1/4$ if they are antiparallel:

$$S_i^z S_j^z = +1/4 \quad \text{for parallel spins}$$
$$S_i^z S_j^z = -1/4 \quad \text{for antiparallel spins.}$$

It then follows that

$$2S_i^z S_j^z - 1/2 = 0 \quad \text{for parallel spins}$$
$$= -1 \quad \text{for antiparallel spins}$$

so that the quantity $-(2S_i^z S_j^z - 1/2)$ can be used to count the number of antiparallel spins. Similarly

$$2S_i^z S_j^z - 1/2 = 1 \quad \text{for parallel spins}$$
$$= 0 \quad \text{for antiparallel spins}$$

so in a similar way the quantity $2S_i^z S_j^z + 1/2$ can be used to count the number of parallel spins.

Thus we have

$$N_{\uparrow\downarrow} = -\sum_{i>j\,\text{nn}} \left(2S_i^z S_j^z - 1/2\right)$$

$$N_{\uparrow\uparrow} = \sum_{i>j\,\text{nn}} \left(2S_i^z S_j^z + 1/2\right)$$

where $\uparrow\uparrow$ indicates parallel spins of either orientation. In terms of the interaction energies $\varepsilon_{\uparrow\uparrow}$ and $\varepsilon_{\uparrow\downarrow}$ introduced in the previous section the energy of the system is then

$$E = N_{\uparrow\uparrow}\varepsilon_{\uparrow\uparrow} + N_{\uparrow\downarrow}\varepsilon_{\uparrow\downarrow}$$

$$= -2(\varepsilon_{\uparrow\downarrow} - \varepsilon_{\uparrow\uparrow}) \sum_{i>j\,\text{nn}} S_i^z S_j^z + \text{const.}$$

The constant term may be ignored. And then the expression for the Ising Hamiltonian may be written as

$$\mathcal{H} = -2\hbar J \sum_{i>j\,\text{nn}} S_i^z S_j^z$$

where $\hbar J = \varepsilon_{\uparrow\downarrow} - \varepsilon_{\uparrow\uparrow}$. Note that the definition of J in our treatment of the Ising model is a factor of four greater than that used in conventional treatments. This has been done purposely to maintain consistency with the J used in the Heisenberg model.

We can see from the expression for $\mathcal{H}$ this is just a part of the Heisenberg Hamiltonian considered previously. The Heisenberg interaction $\mathbf{S}_i \cdot \mathbf{S}_j$ is here approximated by the product $S_i S_j$ — essentially we are taking only the z–z part of the dot product.

If we include an external magnetic field B parallel to the up direction then the Hamiltonian is

$$\mathcal{H} = -2\hbar J \sum_{i>j\,\mathrm{nn}} S_i^z S_j^z - \gamma \hbar B \sum_i S_i^z.$$

The absence of any x or y spin operators is of vital importance. There are only z spin operators and these will therefore all commute. This means that while we have used the quantum-mechanical language of spin $1/2$ and spin operators this is simply a convenient mathematical device and the model is really classical; there are no quantum aspects to the system.

We should also point out that the Weiss mean field treatment of the Heisenberg ferromagnet is equally appropriate for the Ising model. Thus most of the results in Sec. 4.3 carry over to the Ising case. The only difference is that the order parameter for the Heisenberg magnet can point in any direction while the Ising magnetisation can only point "up" or "down". The Heisenberg magnet breaks a continuous symmetry while the Ising transition breaks a discrete symmetry.

4.4.3. *Ising model in one dimension*

As Ising discovered, in one dimension the model exhibits no phase transition. This is an important deviation from the mean field prediction, indicating that the spatial dimension is important. The absence of a transition in 1d may be seen in the following elegant argument, due to Landau.[11]

The essence of Landau's argument is to show that at any non-zero temperature the ordered phase is energetically unfavourable. Let us assume that there is an ordered phase below some critical temperature T_c and let us examine the stability of this phase. We are considering a chain of N ordered spins. Now let us introduce a small element of disorder by reversing all spins from the nth site onwards.

This will increase the energy by $\hbar J$ (there is no applied magnetic field), the contribution coming from the opposite-pointing pair. The entropy of

Fig. 4.21.   Chain with spins reversed from one point onwards.

the system will also increase. Since there is a choice of N sites, after which spins are flipped, the reversal will increase the entropy by $k \ln N$. The effect of the reversal is thus to increase the free energy $F = E - TS$ by

$$\Delta F = \hbar J - kT \ln N.$$

The equilibrium state of the system is that for which F is a minimum. In the thermodynamic limit ($N \to \infty$) the second term of ΔF will always dominate so that the reversal operation results in a reduction of the free energy. And further reversal operations will further reduce the free energy. Thus for all non-zero temperatures the entropy term triumphs over the energy term and the equilibrium configuration for the system is the disordered state. So only at $T = 0$ can the ordered state exist; at all finite temperatures it is unstable. There is no finite-temperature phase transition in an infinite one-dimensional lattice.

This behaviour is shown from a full microscopic calculation. In the presence of a magnetic field B the fractional magnetisation of the one-dimensional Ising system is given (see Plische and Bergersen[12] for details) by

$$m = \frac{\sinh \gamma \hbar B/kT}{\left(\sinh^2 \gamma \hbar B/kT + e^{-\hbar J/kT} \right)^{1/2}}.$$

As expected, we see that when $B = 0$ there is no spontaneous magnetisation at any temperature except $T = 0$. Nevertheless at low temperatures such that

$$\sinh^2 \gamma \hbar B/kT \gg e^{-\hbar J/kT},$$

for any field B, however small, there will occur the saturation magnetisation.

4.4.4.   *Ising model in two dimensions*

The literature on the 2d Ising model is vast; there are even entire books devoted to the subject. There are various methods for calculating the partition function and the transition temperature. However all are complicated, requiring more space than we can reasonably devote in this book. We thus content ourselves with quoting the main results. For further details, we refer the reader to Landau and Lifshitz.[11]

In two dimensions the Ising model exhibits a phase transition. The analytic expression for the transition temperature was derived by Onsager in 1944. The free energy was found, somewhat later, to be

$$F = -Nk\ln(2\cosh \hbar J/2kT) - \frac{Nk}{2\pi} \int_0^{\pi} \ln \frac{1}{2}\left(1 + \sqrt{1 - \kappa^2 \sin^2 \varphi}\right) d\varphi$$

where

$$\kappa = \frac{2}{\cosh \hbar J/2kT \coth \hbar J/2kT}.$$

The transition occurs at the singularity in F, at a critical temperature of

$$T_c = 0.567\hbar J/k,$$

the solution of the equation

$$2\tanh^2 \hbar J/2kT_c = 1 \quad \text{or} \quad \sinh \hbar J/2kT_c = 1.$$

The spontaneous (reduced) magnetisation, the order parameter, is given by

$$m = \{1 - (\sinh \hbar J/2kT)^{-4}\}^{1/8}.$$

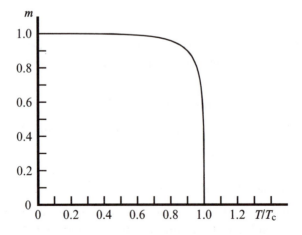

Fig. 4.22.   Order parameter for 2d Ising model.

The critical exponents for the 2d Ising model may be calculated. We compare these with the two we obtained from the mean field model:

2d Ising model	β	γ
mean field	$1/2$	1
exact calc.	$1/8$	$7/4$

The thermal capacity of the Ising system in the vicinity of the phase transition is shown in Fig. 4.23, together with that calculated from two approximate models.

The mean field calculation gives the simplest behaviour. Note that as the temperature is reduced from above the transition there is no "precursor" of the impending critical behaviour. The Bethe–Peirls approximation is an attempt to better the mean field calculation. It considers the effects of nearest neighbour interactions exactly, while the spins further away are treated in mean field. Here the predicted critical temperature is somewhat better and there is a small increase in the thermal capacity before the transition occurs.

The precursor in the thermal capacity is an indication that something is "going on" before the transition actually occurs. This is evidence of the importance of fluctuations in the vicinity of the critical point. In Sec. 4.5.2 we shall see how the mean field/Landau approach indicates, through the dramatic flattening of the free energy at the critical point, that large fluctuations in the order parameter can occur at negligible cost. However mean field theories takes no account of the existence of such fluctuations. The initial assumption that all particles experience the same constant "mean field", precludes the incorporation of such fluctuations.

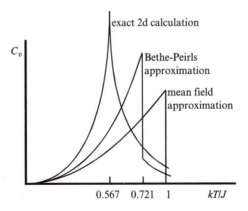

Fig. 4.23.   Heat capacity of 2d Ising model.

4.4.5. *Mean field critical exponents*

In treating the Weiss model of the ferromagnet we obtained values for two of the critical exponents. In that mean field treatment no mention was made of the dimensionality of the order parameter or of the dimensionality of space. Thus mean field arguments will apply equally to the ferromagnet and the Ising model. Note, however, that in contrast to the results of the above two sections, the mean field treatment gives results independent of the spatial dimension.

Recall we found $\beta = 1/2$ for the order parameter exponent and $\gamma = 1$ for the susceptibility exponent. In this section we will use general mean field arguments to extend that discussion, to obtain α, the thermal capacity exponent and δ the equation of state exponent.

(a) Thermal capacity critical exponent α

The thermal capacity is found from the internal energy of the system by differentiating with respect to temperature. Now the internal energy is given by

$$E = -MB$$

where B is the sum of the external and internal fields.

So in the absence of an external field we have

$$\begin{aligned} E &\sim M^2 \quad T < T_c \\ E &= 0 \qquad T > T_c. \end{aligned}$$

But we know that $M \sim (T - T_c)^{1/2}$; that is the order parameter critical behaviour. Thus, we have

$$\begin{aligned} E &\sim T - T_c \quad T < T_c \\ E &= 0 \qquad\quad T > T_c. \end{aligned}$$

And differentiating to find the thermal capacity, this gives

$$\begin{aligned} C &= \text{const} \quad T < T_c \\ &= 0 \qquad\quad T > T_c. \end{aligned}$$

The result is that the thermal capacity is *discontinuous* at the critical point. Observe this in the mean field curve of Fig. 4.23. Such behaviour is observed in the superconducting transition, but not for real ferromagnets.

In terms of a critical exponent, a discontinuity is understood as a power law behaviour

$$C \sim \frac{1}{x}(T - T_c)^x$$

as the quantity x tends to zero. Thus we have the conclusion that the thermal capacity critical exponent α is zero in the mean field model.

(b) Equation of state critical exponent δ

We now turn to the critical exponent connected with the equation of state at $T = T_c$. Recall that δ is defined by

$$M \sim B^{1/\delta}$$

evaluated at the critical point. We have to consider here the behaviour of the magnet in the presence of a magnetic field. The equation of state is conveniently written as

$$\frac{M}{M_0} = \tanh\left\{ \frac{M_0 B}{NkT} + \frac{M}{M_0}\frac{T_c}{T} \right\}.$$

Now in the vicinity of the critical point M is small and if we restrict the discussion to small values of B then we may expand tanh. Thus we expand as far as

$$\frac{M}{M_0} = \frac{M_0 B}{NkT} + \frac{M}{M_0}\frac{T_c}{T} - \frac{1}{3}\left\{ \frac{M_0 B}{NkT} + \frac{M}{M_0}\frac{T_c}{T} \right\}^3 + \cdots.$$

Since we are considering the relation between M and B *at* the critical point, we set $T = T_c$, giving

$$\frac{M}{M_0} = \frac{M_0 B}{NkT_c} + \frac{M}{M_0} - \frac{1}{3}\left\{ \frac{M_0 B}{NkT_c} + \frac{M}{M_0} \right\}^3,$$

or

$$\frac{M_0 B}{NkT_c} = \frac{1}{3}\left\{ \frac{M_0 B}{NkT_c} + \frac{M}{M_0} \right\}^3.$$

This can be rearranged as

$$\frac{M}{M_0} = \left\{ \frac{3M_0 B}{NkT_c} \right\}^{1/3} - \frac{M_0 B}{NkT_c}.$$

Now when B is small it is the first term on the right-hand side that will dominate. Thus we find that at the critical point

$$M \sim B^{1/3}$$

and so we identify the critical exponent δ as 3 in the mean field approximation.

We are unable to derive the mean field results $\nu = {}^1\!/_2$ and $\eta = 0$ here as the calculations are rather more complicated. See Huang[2] for details.

We give below the critical exponents for the three-dimensional ferromagnet, comparing values calculated from the mean field theory with those measured experimentally.[13]

3d ferromagnet	α	β	γ	δ	ν	η
mean field	0	${}^1\!/_2$	1	3	${}^1\!/_2$	0
experiment	−0.14	0.3	1.4		0.7	0.04

In the case of the two-dimensional Ising model we can compare the true calculated critical exponents with those from the mean field approximation.

2d Ising	α	β	γ	δ	ν	η
mean field	0	${}^1\!/_2$	1	3	${}^1\!/_2$	0
exact calculation	0	${}^1\!/_8$	${}^7\!/_4$	15	1	${}^1\!/_4$

From these results we must conclude that while the mean field theory provides a good *qualitative* model for behaviour at the critical point, quantitatively it leaves something to be desired. This is because of the limitations imposed by the fundamental assumption of the mean field model; using a *mean* field involved neglecting the variation in the actual field from site to site. It is precisely these fluctuations, which are neglected, that determine the behaviour in the vicinity of the critical point.

4.4.6.   *The XY model*

The interaction Hamiltonian for the XY model is given by

$$\mathcal{H} = -2\hbar J \sum_{i>j\,\mathrm{nn}} \left(S_i^x S_j^x + S_i^y S_j^y \right).$$

This is an extension of the Ising Hamiltonian, but not the full Heisenberg Hamiltonian. The order parameter for this model is the magnetisation in the x–y plane; it is a vector of dimension $D = 2$. Thus the order parameter is two-dimensional. Since the magnetisation can point in any direction in the x–y plane, it can vary continuously; the XY transition thus breaks a continuous symmetry. This model is regarded as a good description for the superfluid transition in liquid ^{4}He. Estimates for critical exponents for this model in three spatial dimensions are $\gamma = {}^4\!/_3$ and $\alpha = 0$.

4.4.7. *The spherical model*

This model was proposed by Mark Kac (pronounced Cats). The spin variables, and the order parameter are here of *infinite* dimension. In practice, one would consider spin variables of dimension D and then take the limit of $D \to \infty$. The interaction Hamiltonian thus takes the form

$$\mathcal{H} = -2\hbar J \sum_{i>j \text{ nn}} \left(S_i^\alpha S_j^\alpha + S_i^\beta S_j^\beta + S_i^\gamma S_j^\gamma + \cdots + S_i^D S_j^D \right).$$

This is a mathematical extension of the Heisenberg model; the spherical model has the disadvantage that it does not really correspond to a physical system. But in 1952, Theodore Berlin obtained the solution of this model in one, two and three dimensions. He found that in one and two dimensions there was no transition to an ordered phase, but there was a transition at finite temperature in three dimensions. For a three-dimensional simple cubic lattice the critical temperature is found to be

$$kT_c = 0.989\hbar J.$$

So unlike the Ising model, there is no transition in two dimensions.

4.5. Landau Treatment of Phase Transitions

4.5.1. *Landau free energy*

In order to develop a general theory of phase transitions it is necessary to extend the concept of the free energy. For definiteness we will, in this section, consider the case of a ferromagnet, but it should be appreciated that the ideas introduced apply more generally.

The (magnetic) Helmholtz free energy has proper variables T and B. In differential form

$$dF = -S\,dT - M\,dB$$

and the entropy and magnetisation are thus given by

$$S = -\frac{\partial F}{\partial T}\bigg|_B, \quad M = -\frac{\partial F}{\partial B}\bigg|_T.$$

When the temperature and magnetic field are specified the magnetisation (the order parameter in this system) is determined — from the second of the above differential relations. It is possible however, perhaps because of some constraint, that the system may be in a *quasi-equilibrium* state with a different value for its magnetisation. Then the Helmholtz free energy corresponding to that state will be greater than its equilibrium value; it will be a minimum when M takes its equilibrium value. This is equivalent to the law of entropy increase discussed in Chapter 1 and extended in Appendix 2 on thermodynamic potentials.

The conventional Helmholtz free energy applies to systems in thermal equilibrium, as do all thermodynamic potentials. The above discussion leads us to the introduction of a "constrained" Helmholtz free energy for quasi-equilibrium states, which we shall write as

$$F(T, B : M).$$

The conventional Helmholtz free energy is a mathematical function of T and B; when T and B are specified, then the equilibrium state of the system has Helmholtz free energy $F(T, B)$. The system we are considering here is prevented from achieving its full equilibrium state since its magnetisation is constrained to take a certain value. The full equilibrium state of the system is that for which $F(T, B : M)$ is minimised with respect to variations in M, i.e.

$$\frac{\partial F(T, B : M)}{\partial M} = 0.$$

This constrained free energy is often called the Landau free energy (or *clamped* free energy). In a "normal" system one would expect the Landau free energy to possess a simple minimum at the equilibrium point, as shown in Fig. 4.24.

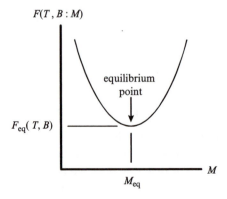

Fig. 4.24.   Variation of Landau free energy with magnetisation.

4.5.2. *Landau free energy for the ferromagnet*

In order to visualise what underlies the Landau theory of phase transitions we shall review the Weiss model of the ferromagnetic transition from the point of view of constrained free energy. Thus we shall consider the Landau free energy of this system.

The Helmholtz free energy is defined as

$$F = E - TS.$$

We will evaluate the internal energy and the entropy separately. Since we require the *constrained* free energy we must be sure to keep M as an explicit variable. The internal energy is given by

$$E = - \int B dM.$$

The magnetic field, in the Weiss model, is the sum of the applied field and the local (mean) field

$$B = B_0 + b.$$

We shall write the local field in terms of the critical temperature:

$$b = \frac{Nk}{M_0^2} T_c M.$$

Integrating up the internal energy we obtain

$$E = -B_0 M - \frac{NkT_c}{2} \left(\frac{M}{M_0} \right)^2.$$

For the present we will consider the case where there is no external applied field. Then $B_0 = 0$, and in terms of the reduced magnetisation $m = M/M_0$ (the order parameter) the internal energy is

$$E = -\frac{NkT_c}{2} m^2.$$

Now we turn to the entropy. This is most easily obtained from the definition

$$S = -Nk \sum_j p_j \ln p_j$$

where p_j are the probabilities of the single-particle states. It is simplest to treat spin $\frac{1}{2}$, which is appropriate for electrons. Then there are two states to sum over:

$$S = -Nk\left[p_\uparrow \ln p_\uparrow + p_\downarrow \ln p_\downarrow\right].$$

Now these probabilities are simply expressed in terms of m, the fractional magnetisation

$$p_\uparrow = \frac{1+m}{2} \quad \text{and} \quad p_\downarrow = \frac{1-m}{2}$$

so that the entropy becomes

$$S = \frac{Nk}{2}[2\ln 2 - (1+m)\ln(1+m) - (1-m)\ln(1-m)].$$

We now assemble the free energy $F = E - TS$, to obtain

$$F = -\frac{Nk}{2}\{T_c m^2 + T[2\ln 2 - (1+m)\ln(1+m) - (1-m)\ln(1-m)]\}.$$

This is plotted for temperatures less than, equal to and greater than the critical temperature.

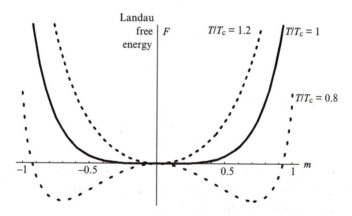

Fig. 4.25.   Landau free energy for Weiss model ferromagnet.

The occurrence of the ferromagnetic phase transition can be seen quite clearly from this figure. For temperatures above T_c we see there is a single minimum in the Landau free energy at $m = 0$, while for temperatures below T_c there are two minima on either side of the origin. The symmetry changes precisely at T_c. There the free energy has flattened, meaning that m may make excursions around $m = 0$ with negligible cost of free energy — hence the large fluctuations at the critical point.

Strictly speaking, Fig. 4.25 applies to the Ising case where the order parameter is a scalar; the order parameter must choose between the values $\pm m$. The XY model is better represented by Fig. 4.26 where, below the transition, the magnetisation can point anywhere in the x–y plane in the minimum of the "wine bottle" potential. Similarly for the 3d Heisenberg model, Fig. 4.26 would apply where m_x and m_y axes really refer to the three axes of the order parameter m_x, m_y and m_z.

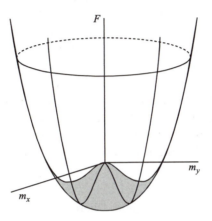

Fig. 4.26.   Low temperature Landau free energy.

The behaviour in the vicinity of the critical point is even clearer if we expand the free energy in powers of the order parameter. Straightforward evaluation gives, neglecting the constant term,

$$F = \frac{Nk}{2}\left\{(T - T_c)m^2 + \frac{T_c}{6}m^4 + \cdots\right\}.$$

We see that the essential information is contained in the *quartic* form for the free energy in the vicinity of the critical point. This will have either one or two minima, depending on the values of the coefficients; the critical point occurs where the coefficient of m^2 changes sign. This abstraction of the *essence* of critical phenomena in terms of the general behaviour of the free energy was formalised by Landau. We have replaced the variable T in the second term by the constant T_c since we are restricting discussion to the vicinity of the critical point.

4.5.3. *Landau theory — second-order transitions*

Landau's theory concerns itself with what is happening in the *vicinity* of the phase transition. In this region the magnitude of the order parameter will be small and the temperature will be close to the transition temperature. When the problem is stated in this way the approach is clear: the Landau free energy should be expanded as a power series in these small parameters. The fundamental assumption of the Landau theory of phase transitions is that in the vicinity of the critical point the free energy is an *analytic* function of the order parameter so that it may indeed be thus expanded. In fact this assumption is, in general, invalid. The Landau theory is equivalent to mean field theory (in the vicinity of the transition) and, as with mean field theory, it is the fluctuations in the order parameter at the transition point which are neglected in this approach. Nevertheless, Landau theory provides a good insight to the general features of phase transitions, describing many of the qualitative features of systems even if the values of the critical exponents are not quite correct.

The free energy is written as a power series in the order parameter φ

$$F = F_0 + F_1\varphi + F_2\varphi^2 + F_3\varphi^3 + F_4\varphi^4 + \cdots$$

First we note that F_0 can be ignored here since the origin of the energy is entirely arbitrary. Considerations of symmetry and the fact that F is a scalar may determine that other terms should be discarded. For instance, if the order parameter is a vector (magnetisation) then odd terms must be discarded since only $\mathbf{M} \cdot \mathbf{M}$ will give a scalar. In this case, the free energy is symmetric in φ, so that

$$F = F_2\varphi^2 + F_4\varphi^4 + \cdots .$$

There is a minimum in F when $\mathrm{d}F/\mathrm{d}\varphi = 0$. Thus the equilibrium state may be found from

$$\frac{\mathrm{d}F}{\mathrm{d}\varphi} = 2F_2\varphi + 4F_4\varphi^3 = 0.$$

This has solutions

$$\varphi = 0 \quad \text{and} \quad \varphi = \pm\sqrt{\frac{-F_2}{2F_4}}.$$

We must have F_4 positive so that F increases far away from the minima to ensure stability. Then if $F_2 < 0$ there are three stationary points while if

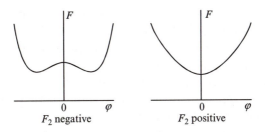

Fig. 4.27.   Minima in free energy for positive and negative F_2.

$F_2 > 0$ there is only one. Thus the nature of the solution depends crucially on the sign of F_2 as can be seen in Fig. 4.27.

It is clear from the figures that the critical point corresponds to the point where F_2 changes sign. Thus expanding F_2 and F_4 in powers of $T - T_c$ we require, to leading order

$$F_2 = a(T - T_c) \quad \text{and} \quad F_4 = b$$

so that

$$\varphi = 0 \qquad\qquad T > T_c$$

$$\varphi = \pm\sqrt{\frac{a(T_c - T)}{2b}} \quad T < T_c.$$

This behaviour reproduces the critical behaviour of the order parameter, as previously calculated for the Weiss model.

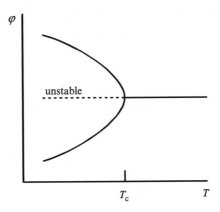

Fig. 4.28.   Behaviour of the order parameter.

We see that when cooling through the critical point the equilibrium magnetisation starts to grow. But precisely at the transition point the magnetisation must decide in which way it is going to point. This is a *bifurcation point*. It is referred to as a pitchfork bifurcation because of its shape.

This model gives the critical exponent for the order parameter, β, to be $1/2$. This is expected, since in the vicinity of the critical point the description is equivalent to the Weiss model.

When φ is small (close to the transition) it follows that the leading terms in the expansion are the dominant ones and we can ignore the higher order terms. We have seen in this section that only terms up to fourth-order in φ were necessary to give a Landau free energy leading to a second-order transition: a free energy that can vary between having a single and a double minimum depending on the values of the coefficients. Landau's key insight here was to appreciate that it is not that the higher order terms *may* be discarded, it is that they *must* be discarded in order to exhibit the *generic* properties of the transition. When terms above φ^4 are discarded this is known as the φ^4 model.

4.5.4. *Thermal capacity in the Landau model*

We have already seen that the Weiss model leads to a discontinuity in the heat capacity (critical exponent $\alpha = 0$). We can now examine this from the perspective of the Landau theory. The free energy is given by

$$F = F_0 + a(T - T_c)\varphi^2 + b\varphi^4.$$

We have included the F_0 term here and we shall permit it to have some smooth temperature variation, which is nothing to do with the transition.

The entropy is found by differentiating the free energy

$$S = -\frac{\partial F}{\partial T} = S_0 - a\varphi^2.$$

This shows the way the entropy drops as the ordered phase is entered. We see that the entropy is continuous at the transition

$$T > T_c, \quad \varphi = 0 \qquad\qquad S = S_0(T)$$

$$T < T_c, \quad \varphi = \pm\sqrt{\frac{a(T_c - T)}{2b}} \quad S = S_0(T) + \frac{a^2}{2b}(T - T_c).$$

The thermal capacity is given by

$$C = \frac{\partial Q}{\partial T} = T\frac{\partial S}{\partial T}.$$

Thus we find

$$T > T_c, \quad C = C_0$$
$$T < T_c, \quad C = C_0 + \frac{a^2 T}{2b}.$$

At the transition there is a discontinuity in the thermal capacity given by

$$\Delta C = \frac{a^2 T_c}{2b}.$$

This is in accord with the discussion of the Weiss model; here we find the magnitude of the discontinuity in terms of the Landau parameters.

4.5.5. *Ferromagnet in a magnetic field*

In the presence of a magnetic field a ferromagnet exhibits certain features characteristic of a first-order transition. There is no latent heat involved, but the transition can exhibit hysteresis. The literature is divided as to whether this is truly a first- or second-order transition. In the case of a first-order transition the free energy may no longer be symmetric in the order parameter; odd powers of φ might appear in F. And the energy of a ferromagnet in the presence of a magnetic field has an additional term $(-\mathbf{M} \cdot \mathbf{B})$ linear in the magnetisation.

For temperatures above T_c the effect of a free energy term linear in φ is to shift the position of the minimum to the left or right. Thus there is an equilibrium magnetisation proportional to the applied magnetic field.

Let us consider the Ising magnet in the presence of a magnetic field. For temperatures below T_c the effect of the linear term is to raise one minimum and to lower the other. We plot the form of F as a function of the order parameter for a number of different applied fields.

We see that in zero external field, since F is symmetric in φ, there are two possible values for the order parameter on either side of the origin. Once a field is applied the symmetry is broken. Then one of the states is the absolute minimum. But the system will remain in the metastable equilibrium. As the field is increased the left-hand side of the right-hand minimum gets shallower and shallower. When it flattens it is then possible for the system to move to the absolute minimum on the other side.

These considerations indicate that the behaviour of the magnetisation discussed in Sec. 4.3.1 is not quite correct for the Ising magnet. There, we stated that the magnetisation inverts as B goes through zero. But in reality there will be *hysteresis*, a specific characteristic of first-order transitions.

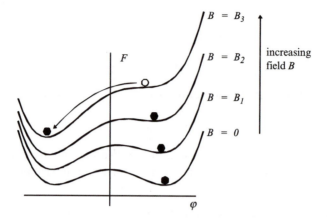

Fig. 4.29.   Variation in free energy for different magnetic fields.

The discussion gave the equilibrium behaviour whereas hysteresis is a quasi-equilibrium phenomenon.

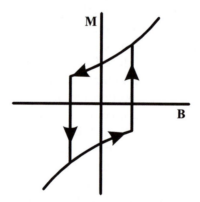

Fig. 4.30.   Hysteretic variation of magnetisation below T_c.

This hysteretic behaviour we have discussed only applies to the Ising magnet, not the XY magnet nor the Heisenberg magnet which both have broken *continuous* symmetries. Then the magnetisation can always shift in the distorted "wine bottle" potential to move to the lowest free energy.

Real (Heisenberg) magnets do, however, exhibit hysteresis. This happens through the formation of *domains*: regions where the magnetisation points in different directions. Domain formation provides extra entropy from the domain walls at some energy cost. This can lower the free energy at finite temperatures. When one encounters unmagnetised iron at room temperatures, which is way below its critical temperature of 1044 K, this is because of the cancellation of the magnetisation of the domains.

4.6. Ferroelectricity

4.6.1. *Description of the phenomenon*

In certain ionic solids a spontaneous electric polarisation can appear below a particular temperature. A typical example of this is barium titanate, which has a transition temperature of about 140°C. There is a good description of ferroelectricity in Kittel's *Introduction to Solid State Physics* book.[14]

The essential feature of the ferromagnetic transition is that the "centre of mass" of the positive charge becomes displaced from that of the negative charge. In this way a macroscopic polarisation appears. The polarised state is called a *ferroelectric*. The transition is associated with a structural change in the solid.

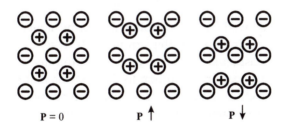

Fig. 4.31.   Unpolarised state and ferroelectric state showing two polarisations.

In describing this transition we see that the order parameter is the electric polarisation **P**. However the polarisation is restricted in the directions it may point. In the figure, it may point either up or down. Thus it is a discrete symmetry that is broken; in reality, the effective order parameter is a scalar. We note that this system has a non-conserved vector order parameter. At the phenomenological level the transition has many things in common with the Ising model of the magnetic transition of the ferromagnet although the description is very different at the microscopic level. One special feature of the ferroelectric case is that the transition may be first- or second-order. We will use the Landau approach to phase transitions to examine this system and we shall be particularly interested in what determines the order of the transition.

Our previous example of a first-order transition, the liquid–gas system was one with a conserved order parameter. Similarly, phase separation in a binary alloy, to be treated in Sec. 4.7 is a first-order system with a conserved order parameter. Here, however, we will encounter a first-order transition where the order parameter is not conserved.

4.6.2. *Landau free energy*

The order parameter is the electric polarisation, but in keeping with our general approach we shall denote it by φ in the following. In the spirit of Landau we shall expand the (appropriate) free energy in powers of φ:

$$F = F_0 + F_1\varphi + F_2\varphi^2 + F_3\varphi^3 + F_4\varphi^4 + F_5\varphi^5 + F_6\varphi^6 + \cdots.$$

On the assumption that the crystal lattice has a centre of symmetry the odd terms of the expansion will vanish. Furthermore in the absence of an applied electric field there will be no $F_1\varphi$ term. We can ignore the constant term F_0 and so the free energy simplifies to

$$F = F_2\varphi^2 + F_4\varphi^4 + F_6\varphi^6 + \cdots.$$

Here, we have allowed for the possibility that terms higher than the fourth power may be required (unlike the ferromagnetic case).

When we considered the ferromagnet we truncated the series at the fourth power. A necessary condition to do this was that the final coefficient, F_4, was positive so that the order parameter remained bounded. If we were to have a system where F_4 was negative then we would have to include higher order terms until a positive one were found. The simplest example is where F_6 is positive and we would truncate there. We will see that in this case the *order* of the transition is determined by the *sign* of the F_4 term.

If F_4 is positive then in the spirit of the Taylor expansion we can ignore the F_6 term and the general behaviour, in the vicinity of the transition, is just as the ferromagnet considered previously. As the temperature is cooled below the transition point the order parameter grows continuously from its zero value.

If F_4 is negative then in the simplest case we need a positive F_6 term to terminate the series. The general form of the free energy curve then has three minima, one at $\varphi = 0$ and one equally spaced on either side.

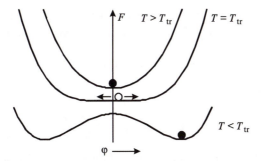

Fig. 4.32. Conventional second-order transition.

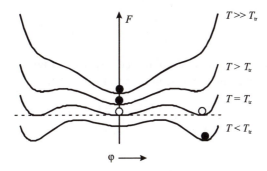

Fig. 4.33. Inclusion of sixth-order terms.

At high temperatures the minimum at $\varphi = 0$ is lower while at low temperatures the minima on either side will be lower. The transition point corresponds to the case where all three minima occur at the same value of F. There is thus a jump in the order parameter at the transition as φ moves from one minimum to another.

Here, we see that the transition is characterised by a jump in the order parameter and there is the possibility of hysteresis in the transition since the barrier to the lower minimum must be surmounted. These are the characteristics of a first-order transition.

4.6.3. *Second-order case*

The analysis follows that previously carried out for the ferromagnet case. Here F_4 is positive and the free energy is

$$F = F_2\varphi^2 + F_4\varphi^4$$

when we ignore the sixth- and higher order terms. There is a minimum in F when $dF/d\varphi = 0$. Thus the equilibrium state may be found from

$$\frac{dF}{d\varphi} = 2F_2\varphi + 4F_4\varphi^3 = 0.$$

This has solutions

$$\varphi = 0 \quad \text{and} \quad \varphi = \pm\sqrt{\frac{-F_2}{2F_4}}.$$

We recall that the nature of the solution depends crucially on the sign of F_2.

[If we include the sixth-order term in the free energy expansion then the non-zero solution for the order parameter is

$$\varphi^2 = -\frac{F_4}{3F_6}\left\{1 - \sqrt{1 - \frac{3F_2F_6}{F_4^2}}\right\}.$$

This reduces to the previously obtained expression when F_6 is neglected, as can be seen from the expansion of φ in terms of F_6:

$$\varphi = \pm\sqrt{\frac{-F_2}{2F_4}}\left\{1 + \frac{3}{8}\frac{F_2F_6}{F_4^2} + \frac{63}{128}\left(\frac{F_2F_6}{F_4^2}\right)^2 + \cdots\right\}.$$

]

We know that F_4 must be positive and the critical point corresponds to the point where F_2 changes sign. Thus expanding F_2 and F_4 in powers of $T - T_c$ we require, to leading order

$$F_2 = a(T - T_c) \quad \text{and} \quad F_4 = b$$

so that

$$\varphi = 0 \qquad\qquad\qquad T > T_c$$

$$\varphi = \sqrt{\frac{a(T_c - T)}{2b}} \quad T < T_c.$$

This is the conventional Landau result for a second-order transition.

4.6.4. *First-order case*

In a first-order transition we *must* include the sixth-order term in the free energy expansion. In this case, we have

$$F = F_2\varphi^2 + F_4\varphi^4 + F_6\varphi^6$$

where we note that $F_4 < 0$ and $F_6 > 0$.

At the transition point there are three minima of the free energy curve and they have the same value of F, namely zero (since we have taken F_0 to be zero). Thus the transition point is characterised by the conditions

$$F(\varphi) = 0 \quad \text{and} \quad \frac{\mathrm{d}F}{\mathrm{d}\varphi} = 0$$

or

$$F_2\varphi^2 + F_4\varphi^4 + F_6\varphi^6 = 0$$
$$2F_2\varphi + 4F_4\varphi^3 + 6F_6\varphi^5 = 0.$$

We know we have the $\varphi = 0$ solution; this can be factored out. The others are found from solving the simultaneous equations

$$F_2 + F_4\varphi^2 + F_6\varphi^4 = 0$$
$$F_2 + 2F_4\varphi^2 + 3F_6\varphi^4 = 0.$$

The solution to these is

$$\varphi^2 = -\frac{F_4}{2F_6}$$

or

$$\varphi = \sqrt{\frac{-F_4}{2F_6}}$$

as we know that F_4 is negative.

This gives the discontinuity in the order parameter at the transition since φ will jump from zero to this value at the transition. Thus we can write

$$\Delta\varphi = \sqrt{\frac{-F_4}{2F_6}}.$$

This vanishes as F_4 goes to zero; in other words, the transition becomes second-order. Thus we have shown that when F_4 changes sign the transition becomes second-order. The point of changeover from first- to second-order is referred to as the *tricritical point*.

At the transition point the value of F_2 may be found as the second solution to the simultaneous equation pair. The result is

$$F_2 = \frac{F_4^2}{4F_6} \neq 0.$$

This means that in the first-order case the transition does *not* correspond to the vanishing of F_2, which we nevertheless continue to call the critical point.

The key argument we use here is that we expect F_2 will continue to vary with temperature in the way of the second-order case:

$$F_2 = a(T - T_c).$$

The value of F_4 may vary with some external parameter such as strain. Then when the transition becomes first-order the transition temperature will increase above its critical value:

$$F_4 > 0, \quad T_{tr} = T_c, \qquad \text{second-order}$$

$$F_4 < 0, \quad T_{tr} = T_c + \frac{1}{4a}\frac{F_4^2}{F_6}, \quad \text{first-order.}$$

This is plotted in Fig. 4.34.

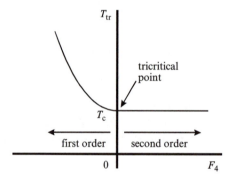

Fig. 4.34.   Variation of transition temperature with F_4.

In the ordered phase the value of the order parameter is that corresponding to the lowest minimum of the free energy. We thus require

$$\frac{\mathrm{d}F}{\mathrm{d}\varphi} = 0,$$

$$\frac{\mathrm{d}^2 F}{\mathrm{d}\varphi^2} > 0$$

where the second condition ensures the stationary point is a minimum. Using the sixth-order expansion for the free energy, differentiating it and discarding the $\varphi = 0$ solution gives

$$\varphi^2 = \frac{|F_4|}{3F_6}\left\{ 1 \pm \sqrt{1 - \frac{3F_2F_6}{F_4^2}} \right\}.$$

Remember that F_4 is negative here. We have to choose the positive root to obtain the *minima*; the negative root gives the maxima. Using the conventional temperature variation for F_2 together with the expression for the transition temperature T_{tr} in terms of the critical temperature T_c gives the temperature variation of the order parameter for the first-order transition:

$$\varphi^2 = \frac{|F_4|}{3F_6}\left\{ 1 + \sqrt{\frac{4T_{tr} - 3T_c - T}{4(T_{tr} - T_c)}} \right\}.$$

The figure below shows the temperature variation of the order parameter for the case where $T_{tr}/T_c = 1.025$.

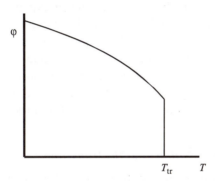

Fig. 4.35. Order parameter in a first-order transition.

This shows clearly the discontinuity in the order parameter at T_{tr}, characteristic of a first-order transition.

Caveat

There is a contradiction in the use of the Landau expansion for treating first-order transitions. The justification for expanding the free energy in powers of the order parameter and the truncation of this expansion is based on the assumption that the order parameter is very small. Another way of expressing this is to say that the Landau expansion is applicable to the vicinity of the critical point. This is fine in the case of second-order transitions where, when the ordered phase is entered, the order parameter grows continuously from zero. However the characteristic of a first-order transition is the discontinuity in the order parameter. In this case, the assumption of a vanishingly small order parameter is invalid. When the ordered phase is entered the order parameter jumps to a finite value. There is then no justification for a small-φ expansion of the free energy. There are two ways of accommodating this difficulty.

Notwithstanding the invalidity of terminating the free energy expansion, the *qualitative* discussion of Sec. 4.6.2 still holds. In particular, it should be apparent that the general features of a first-order transition can be accounted for by a free energy which is a sixth-order polynomial in the order parameter. From this perspective, we may regard the Landau expansion as putting the simplest of mathematical flesh on the bones of a qualitative model.

It is clear, however, that the free energy expansion becomes more respectable as the discontinuity in the order parameter becomes smaller, particularly when it goes to zero. Thus the theory is expected to describe well the changeover between first-order and second-order at the tricritical point. Transitions where the discontinuity in the order parameter is small are referred to as being *weakly first-order*. We conclude that the Landau expansion should be appropriate for weakly first-order transitions.

4.6.5. *Entropy and latent heat at the transition*

The temperature dependence of the free energy in the vicinity of the transition is given by

$$F = F_0(T) + a(T - T_c)\varphi^2 + F_4\varphi^4 + F_6\varphi^6.$$

Here $F_0(T)$ gives the temperature dependence in the disordered, high temperature phase, and the coefficient of φ^2 gives the dominant part of the temperature dependence in the vicinity of the transition. The temperature dependence here is similar to that of the second-order case, treated in the context of the ferromagnet.

The entropy is given by the temperature derivative of F according to the standard thermodynamic prescription

$$S = -\frac{\partial F}{\partial T},$$

in this case

$$S = -\frac{\partial F_0}{\partial T} - a\varphi^2.$$

The second term indicates the reduction in entropy that follows as the order grows in the ordered phase. We see that if there is a discontinuity in the order parameter then there will be a corresponding discontinuity in the entropy of the system.

In the first-order transition there is a discontinuity in the order parameter at $T = T_{tr}$. The latent heat is then given by

$$L = -T_{tr}\Delta S$$
$$= aT_{tr}\Delta\varphi^2.$$

This shows how the latent heat is directly related to the discontinuity in the order parameter — both characteristics of a first-order transition. The discontinuity in φ is given by

$$\Delta\varphi = \sqrt{\frac{-F_4}{2F_6}}$$

so we can write the latent heat as

$$L = aT_{tr}\frac{|F_4|}{2F_6}.$$

F_4 is negative for a first-order transition. This expression shows how the latent heat vanishes as F_4 vanishes when the transition becomes second-order.

4.6.6. *Soft modes*

In the ferroelectric, excitations of the order parameter are optical phonons; the positive and negative charges oscillate in opposite directions. In other words the polarisation oscillates. These are not Goldstone bosons since it is not a continuous symmetry that is broken. And thus the excitations have a finite energy (frequency) in the $p \to 0$ ($k \to 0$) limit. This is a characteristic of optical phonons.

The frequency of the optical phonons depends on the restoring force of the interatomic interaction and the mass of the positive and negative ions. Now, let us consider the destruction of the polarisation as a ferroelectric is warmed through its critical temperature. We are interested here in the case of a second-order transition. At the critical point, the Landau free energy exhibits anomalous broadening; the restoring force vanishes and the crystal becomes unstable. And this means that the frequency of the optical phonon modes will go to zero: they become "soft".

For further details of soft modes see the books by Burns[15] and Kittel[14].

4.7.   Binary Mixtures

We have already seen the gas–liquid transition of a fluid as an example of a system exhibiting a first-order transition with conserved order parameter. And in the previous section the ferroelectric had a first-order transition with non-conserved order parameter. In this section, we shall examine phase separation in a binary mixture. This is a first-order transition with a conserved order parameter. By comparing and contrasting these different systems we should be able to identify their similarities and differences. We should also note that the binary alloy model treated in this section is equivalent to the Ising model (with a conserved order parameter). And the procedures described in Sec. 4.7.2 and beyond correspond to the mean field treatment of this Ising model.

4.7.1.   *Basic ideas*

Consider a mixture of two atomic species A and B, with relative proportions x and $1 - x$. For clarity we shall consider a solid system, but many of the arguments will also apply to liquid mixtures. We thus imagine an alloy whose composition is specified by A_xB_{1-x}. We assume that the energy of the system is specified in terms of the nearest neighbour interactions and we will denote

ε_{aa} energy of a single a—a bond

ε_{bb} energy of a single b—b bond

ε_{ab} energy of a single a—b bond.

A microstate of this system is specified by indicating the occupancy of each lattice site as an A or a B atom. Each microstate will have a given energy, found from a consideration of what neighbours each site has. And from these energies the partition function could be found. This would then give all the thermodynamic properties of the system. Unfortunately, such a procedure would be prohibitively complicated. Instead, we shall use an approximation method.

4.7.2. *Model calculation*

We make the assumption that the A and B atoms are distributed randomly. In other words, we are considering a homogeneous mixture. We are considering a system of specified temperature and volume so we calculate the Landau–Helmholtz free energy for this system. Thus we need to know the energy and the entropy.

The key point is to start by considering a given *bond* joining two neighbouring atoms. We label one atom as the *left* atom and the other as the *right*.

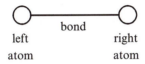

left   bond   right

atom   atom

Fig. 4.36. Bond joining two atoms.

Each of the atoms may be an A atom or a B atom. Thus there are four different configurations for the bond: a—a, a—b, b—a, b—b.

left atom \ right atom	A	B
A	ε_{aa}	ε_{ab}
B	ε_{ab}	ε_{bb}

configuration energies

The concentration of the A atoms is taken to be x; then the concentration of the B atoms is $(1 - x)$. We assume the atoms are distributed randomly so the probability that a site is occupied by an A atom is x. And the probability it is occupied by a B atom is $(1 - x)$. In the general case we could allow the concentration to vary with position. Then the probability the left atom is A is x_l; the probability it is B is $1 - x_l$; the probability the right atom is A is x_r; the probability it is B is $1 - x_r$. The four bond configurations a—a, a—b, b—a, b—b, then have the probabilities: $x_l x_r$, $x_l(1 - x_r)$, $(1 - x_l)x_r$, $(1 - x_l)(1 - x_r)$.

left atom \ right atom	A	B
A	$x_l x_r$	$x_l(1-x_r)$
B	$(1-x_l)x_r$	$(1-x_l)(1-x_r)$

configuration probabilities

4.7.3. *System energy*

We have defined the energies $\{\varepsilon_{aa}, \varepsilon_{bb}, \varepsilon_{cc}\}$ of the three types of bonds above and we now know the probability of occurrence of each configuration. The mean energy for the bond will be the sum of the energies of each state multiplied by its probability

$$\bar{e}_{lr} = \varepsilon_{aa}x_l x_r + \varepsilon_{ab}x_l(1-x_r) + \varepsilon_{ab}(1-x_l)x_r + \varepsilon_{bb}(1-x_l)(1-x_r).$$

We are assuming that the composition of the system is homogeneous so that the concentration is independent of position; $x_l = x_r = $ constant, x. In this case, the mean energy per bond reduces to

$$\bar{e} = x^2 \varepsilon_{aa} + (1-x)^2 \varepsilon_{bb} + 2x(1-x)\varepsilon_{ab}.$$

which may be rearranged as

$$\bar{e} = x\varepsilon_{aa} + (1-x)\varepsilon_{bb} + x(1-x)\{2\varepsilon_{ab} - (\varepsilon_{aa} + \varepsilon_{bb})\}.$$

The expression for the internal energy is found by considering a system containing N atomic sites in a lattice where each atom has s neighbours. Then the number of neighbour bonds will be $Ns/2$; the divisor of 2 removes double counting. The internal energy of the system is then $E = Ns\bar{e}/2$ or

$$E = \frac{Ns}{2}\left[x^2 \varepsilon_{aa} + (1-x)^2 \varepsilon_{bb} + 2x(1-x)\varepsilon_{ab}\right].$$

This may be written in a more suggestive form as

$$E = \frac{Ns}{2}\left[x\varepsilon_{aa} + (1-x)\varepsilon_{bb} + 2x(1-x)\left(\varepsilon_{ab} - \left(\frac{\varepsilon_{aa} + \varepsilon_{bb}}{2}\right)\right)\right]$$

where the first two terms represent the energy of the separated pure phases E_0

$$E_0 = \frac{Ns}{2}[x\varepsilon_{aa} + (1-x)\varepsilon_{bb}],$$

and the third term thus gives the energy of mixing E_m

$$E_m = Nsx(1-x)\left(\varepsilon_{ab} - \left(\frac{\varepsilon_{aa} + \varepsilon_{bb}}{2}\right)\right).$$

It is then convenient to define the energy

$$\varepsilon = \varepsilon_{ab} - \frac{1}{2}(\varepsilon_{aa} + \varepsilon_{bb}),$$

the difference between the "unlike" neighbour energy and the mean of the two "like" neighbour energies. This will turn out to be the characteristic energy of the system. Then, the energy of mixing takes the simple form

$$E_m = Nsx(1-x)\varepsilon.$$

Observe that the energy of mixing is invariant under the transformation $x \rightarrow 1 - x$. In other words E_m is symmetric about the line $x = 1/2$. Any system whose energy satisfies this condition (regardless of the precise expression for its energy of mixing) is referred to as a *strictly regular solution*.

An *ideal solution* is one for which $\varepsilon = 0$. In this case the energy of mixing is zero; the energy is independent of the microscopic structure. An ideal solution cannot lower its energy by changing its structure.

By contrast when $\varepsilon \neq 0$ then the solution can lower its energy by rearranging itself in an ordered structure. When $\varepsilon < 0$ then a—b bonds are preferred and the ordered state will be a superlattice. (Clearly this would only happen in a solid.) And when $\varepsilon > 0$ then a—a and b—b bonds are preferred; the ordered state will comprise separate regions of phase-separated atoms. (This would happen in both solids and liquids.)

4.7.4. *Entropy*

Each site can be occupied by an A atom or a B atom; it has two states. The probability that it is an A atom is x and the probability that it is a B atom is $(1-x)$. The entropy of a given site is then $-k[x \ln x + (1-x)\ln(1-x)]$, so that for a solid of N atomic sites the entropy will be

$$S = -Nk[x \ln x + (1-x)\ln(1-x)].$$

You should recognise that this expression for the entropy is identical to that for the spin $1/2$ magnet; there each site also had two states.

4.7.5. *Free energy*

The system has its temperature and volume fixed thus the Helmholtz free energy is the appropriate thermodynamic potential to use. Since

$$F = E - TS$$

we then have

$$F = \frac{Ns}{2}[x\varepsilon_{aa} + (1-x)\varepsilon_{bb} + 2x(1-x)\varepsilon] + NkT[x\ln x + (1-x)\ln(1-x)].$$

This is plotted in Fig. 4.37. We see that at high temperatures the free energy curve has a single minimum whereas at lower temperatures a double minimum develops.

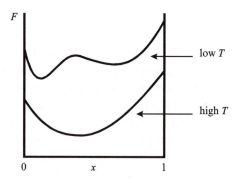

Fig. 4.37.   Free energy of binary alloy.

We shall see that the low temperature free energy leads to phase separation.

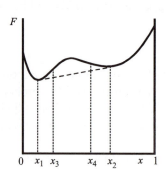

Fig. 4.38.   Free energy leading to phase separation.

Between $x = x_1$ and $x = x_2$ it is possible to lower the free energy by phase separation by dropping down from the curve to the double tangent line, with a mixture of phases at densities x_1 and x_2. This is analogous to the situation with the liquid–gas system treated in the van der Waals approach in Sec. 4.2.1. And as with the fluid case, in the phase separation region some parts are metastable while some parts are unstable.

Instability occurs whenever $\partial F/\partial x$ is a *decreasing* function of x, that is, when $\partial^2 F/\partial x^2 < 0$. Then the system is unstable with respect to infinitesimal density fluctuations. This happens in the region x_3 to x_4. This is known as the *spinodal* region. If the temperature is quenched into this region then one has spontaneous phase separation, referred to as *spinodal decomposition*.

By contrast, the regions x_1 to x_3 and x_4 to x_2 are *metastable*. Here, it is possible to remain in the inhomogeneous phase unless a density fluctuation of sufficient magnitude occurs. The system is unstable with respect to *finite* density fluctuations. Clearly, if one waits long enough a fluctuation of sufficient magnitude will occur (but it might be a very long wait indeed).

In the phase diagram (the x–T plane) the homogeneous phase and the metastable phase are separated by the phase separation or *binodal* line. The metastable phase and the unstable phase are separated by the spinodal line. We shall see that for the model considered here the locus of these two lines may be calculated.

4.7.6. *Phase separation — the lever rule*

The phase separation curve is found by the double tangent construction, introduced in Sec. 4.2.1; this determines the region where the free energy may be reduced from its "homogeneous phase" value by separating into two phases.

The system will separate into regions of concentration x_1 and regions of concentration x_2. The *fractions* of the substance in the two phases are determined by the *lever rule*. Let us denote the fraction of the substance in the phase of concentration x_1 by α; then the fraction in the other phase, of concentration x_2 will be $1 - \alpha$. The number of A atoms in the x_1 phase will be $N\alpha x_1$, while the number of A atoms in the x_2 phase will be $N(1 - \alpha)x_2$. The sum of these will give the total number of A atoms in the system, Nx_0, where x_0 is the concentration of A atoms in the homogeneous phase. Thus

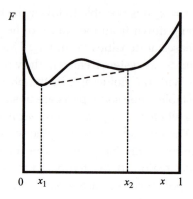

Fig. 4.39.   Double tangent construction.

we have the equality

$$Nx_0 = N\alpha x_1 + N(1 - \alpha)x_2.$$

This may be solved for the fractions α and $1 - \alpha$ to give

$$\alpha = \frac{x_2 - x_0}{x_2 - x_1}$$

$$1 - \alpha = \frac{x_0 - x_1}{x_2 - x_1}.$$

This has a simple interpretation in terms of distances on the phase diagram. This is called the *lever rule*.

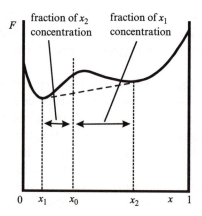

Fig. 4.40.   The lever rule.

4.7.7. *Phase separation curve — the binodal*

The concentrations x_1 and x_2 of the two separated phases are determined from the double tangent construction. So essentially we must solve the simultaneous equations

$$\frac{dF(x)}{dx}\bigg|_{x_1} = \frac{dF(x)}{dx}\bigg|_{x_2}$$

$$F(x_2) = F(x_1) + (x_2 - x_1)\frac{dF(x)}{dx}\bigg|_{x_2, x_1}$$

for x_1 and x_2.

In general, this could be a difficult problem. But there is an important simplification that helps in the case of a strictly regular solution. We can perform the double tangent construction on the free energy of mixing F_m:

$$F_m = E_m - TS.$$

This is possible because the energy E_0 which we have neglected

$$E_0 = \frac{Ns}{2}[x\varepsilon_{aa} + (1-x)\varepsilon_{bb}]$$

is a linear function of x. This means that removing E_0 from the free energy is equivalent to a vertical shear of the free energy graph. The double tangent construction is unchanged; the straight line still touches the free energy curve tangentially at the two points $x = x_1$ and $x = x_2$.

For a strictly regular solution the free energy of mixing is symmetrical about $x = 1/2$. Thus the two minima will be equally spaced on either side

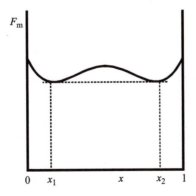

Fig. 4.41. Double tangent construction for free energy of mixing.

of $x = 1/2$, and the symmetry now implies that the minima will be at the same height. In other words, on the free energy of mixing curve the double tangent construction is a *horizontal* line joining the *minima* of the free energy.

The concentrations x_1 and x_2 of the two separated phases are determined from the condition

$$\frac{\mathrm{d}F_{\mathrm{m}}(x)}{\mathrm{d}x} = 0$$

where

$$F_{\mathrm{m}}(x) = Nsx(1 - x)\varepsilon + NkT[x \ln x + (1 - x)\ln(1 - x)].$$

The derivative of this is

$$\frac{\mathrm{d}F_{\mathrm{m}}}{\mathrm{d}x} = Ns\varepsilon(1 - 2x) - NkT \ln\left(\frac{1 - x}{x}\right).$$

We should set this expression equal to zero and then solve the equation for x, giving the two solutions x_1 and x_2 as a function of temperature. Unfortunately an explicit expression cannot be obtained. However, it is possible to express T as a function of x, which gives the locus of the phase separation line:

$$T_{\mathrm{ps}} = \frac{s\varepsilon(1 - 2x)}{k \ln[(1 - x)/x]}.$$

The transformation $x \to 1 - x$ leaves the expression for T_{ps} unchanged, so there are two solutions for x corresponding to a given temperature.

The *critical temperature* T_{c} corresponds to the maximum T_{ps}, occuring at $x = 1/2$ by symmetry. One must be careful in taking the limit correctly. Since

$$\lim_{x \to 1/2} \frac{1 - 2x}{\ln[(1 - x)/x]} = \frac{1}{2}$$

it follows that the critical temperature is given by

$$T_{\mathrm{c}} = \frac{s\varepsilon}{2k}.$$

From this, we can write the phase separation temperature as

$$T_{\mathrm{ps}} = \frac{2(1 - 2x)}{\ln[(1 - x)/x]} T_{\mathrm{c}}.$$

And from this we plot the phase separation curve as shown in Fig. 4.42.

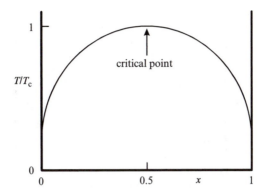

Fig. 4.42.   Phase separation of a binary alloy.

4.7.8. *The spinodal curve*

The spinodal curve traces out the region of instability in the T–x plane. This is characterised by the vanishing of the second derivative of F (or the free energy of mixing), points x_3 and x_4 of Fig. 4.38. Thus, we require

$$\frac{d^2F}{dx^2} = 0.$$

We found the first derivative in the previous section:

$$\frac{dF_m}{dx} = Ns\varepsilon(1-2x) - NkT\ln\left(\frac{1-x}{x}\right)$$

and differentiating once more gives

$$\frac{d^2F_m}{dx^2} = -2Ns\varepsilon + NkT\left(\frac{1}{1-x} + \frac{1}{x}\right).$$

Setting this equal to zero gives the spinodal temperature as a function of x

$$T_{sp} = \frac{2s\varepsilon}{k}x(1-x)$$

or, in terms of the critical temperature

$$T_{sp} = 4x(1-x)T_c.$$

The spinodal line is below the phase separation line. They meet at the critical point where $x = 1/2$ (for a strictly regular solution).

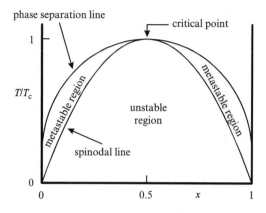

Fig. 4.43.   Phase separation and spinodal line.

4.7.9.   *Entropy in the ordered phase*

In the homogeneous phase we saw that the entropy of the system is given by

$$S = -Nk[x \ln x + (1 - x) \ln(1 - x)]$$

where x is the concentration of the A species. Thus in the homogeneous phase S is independent of T so that the thermal capacity is zero

$$C = 0 \quad \text{in the homogeneous phase.}$$

In the ordered state we have regions of concentration x_1 and regions of concentration x_2. The lever rule tells us how much of each phase we have

$$\text{fraction } \frac{x_2 - x_0}{x_2 - x_1} \text{ of phase 1,}$$

$$\text{fraction } \frac{x_0 - x_1}{x_2 - x_1} \text{ of phase 2.}$$

The entropy of the two phases is

$$S_1 = -N_1 k[x_1 \ln x_1 + (1 - x_1) \ln(1 - x_1)]$$
$$S_2 = -N_2 k[x_2 \ln x_2 + (1 - x_2) \ln(1 - x_2)]$$

so that the total system entropy will be

$$S = -Nk \frac{x_2 - x_0}{x_2 - x_1}[x_1 \ln x_1 + (1 - x_1) \ln(1 - x_1)]$$

$$- Nk \frac{x_0 - x_1}{x_2 - x_1}[x_2 \ln x_2 + (1 - x_2) \ln(1 - x_2)].$$

An important simplification follows for strictly regular solutions. In that case, we have symmetry about $x = 1/2$ and

$$x_2 = 1 - x_1.$$

The entropy is then

$$S = -Nk[x_1 \ln x_1 + (1 - x_1) \ln(1 - x_1)].$$

The entropy is independent of the initial concentration x_0 and it depends only on the concentration of the separated states. Since we could have equally used x_2 in the entropy expression it is clear that the total entropy does not depend explicitly on the relative fractions of the two phases.

It further follows that in the separated region the entropy (and therefore the thermal capacity) will be independent of the initial concentration. The temperature at which phase separation occurs will depend on x_0 but once the separated region is entered the entropy (and therefore the thermal capacity) will join a *universal curve*. This happens for a strictly regular solution.

In the ordered phase we find that the entropy depends on temperature since now entropy depends on x_1 (or x_2) and x_1 (or x_2) depends on temperature:

$$S = S(x_1) \leftarrow S = -Nk[x_1 \ln x_1 + (1 - x_1) \ln(1 - x_1)]$$

and

$$x_1 = x_1(T) \leftarrow T = \frac{2(1 - 2x_1)}{\ln[(1 - x_1)/x_1]} T_c.$$

Although an explicit expression for S in terms of T cannot be found, we can make a parametric plot by varying x_1. This is shown in Fig. 4.44.

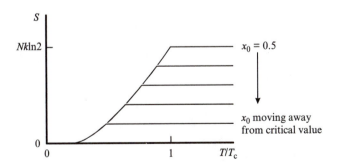

Fig. 4.44.   Entropy of phase separated binary mixture.

4.7.10. *Thermal capacity in the ordered phase*

The heat capacity is found from the entropy

$$C = \frac{dQ}{dT}$$

$$= T\frac{dS}{dT}.$$

The differentiation is performed on a function of a function since entropy depends on x_1, and x_2 depends on T. Thus we use

$$\frac{dS}{dT} = \frac{dS}{dx_1} \Big/ \frac{dT}{dx_1}.$$

The derivatives are evaluated as

$$\frac{dS}{dx_1} = Nk\ln\left(\frac{1-x_1}{x_1}\right)$$

$$\frac{dT}{dx_1} = \frac{2T_c}{\ln[(1-x_1)/x_1]}\left\{\frac{1-2x_1}{x_1(1-x_1)\ln[(1-x_1)/x_1]} - 2\right\}.$$

Again, we cannot assemble these into an *explicit* function of C in terms of T but we can make use of a parametric plot since

$$T = \frac{2(1-2x_1)}{\ln[(1-x_1)/x_1]}T_c.$$

The thermal capacity is

$$S = T\frac{dS}{dT}$$

so that in this case

$$S = \frac{2(1-2x_1)}{\ln\left(\frac{1-x_1}{x_1}\right)}T_c\frac{Nk\ln[(1-x_1)/x_1]}{\frac{2T_c}{\ln[(1-x_1)/x_1]}\left\{\frac{1-2x_1}{x_1(1-x_1)\ln[(1-x_1)/x_1]} - 2\right\}}.$$

Thus, we have the implicit pair

$$\left.\begin{array}{l} C = Nk\dfrac{x_1(1-x_1)(1-2x_1)\ln^2[(1-x_1)/x_1]}{1-2x_1-2(1-x_1)x_1\ln[(1-x_1)/x_1]} \\[4mm] T = \dfrac{2(1-2x_1)}{\ln[(1-x_1)/x_1]}T_c. \end{array}\right\}$$

We plot this in Fig. 4.45.

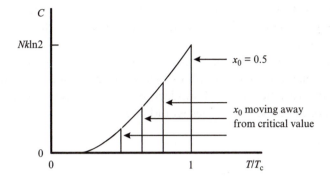

Fig. 4.45. Heat capacity of binary mixture.

If we eliminate $\ln[(1 - x_1)/x_1]$ from these equations then one obtains the expression commonly quoted in the literature (e.g. Slater[16]):

$$\frac{C}{Nk} = \frac{2(1 - 2x_1)^2}{\dfrac{T}{T_c}\left(\dfrac{T/T_c}{2x_1(1 - x_1)} - 2\right)},$$

but this is only one of an implicit pair.

The entropy is continuous at the transition and there is a simple discontinuity in the heat capacity. This means that there is *no latent heat* even though the transition is first-order.

The absence of latent heat is due simply to the constraints under which we are studying the transition. The gas–liquid transition (and indeed the liquid–solid transition) are studied conventionally under the condition of constant pressure, often atmospheric pressure. If the transition were studied at constant volume then there would be no latent heat. The fixed concentration x_0 of the binary mixture is analogous to the fluid system at constant density. To investigate the latent heat of the transition in a binary mixture requires the introduction of the *osmotic pressure* of the components and consideration of the transition under the (hypothetical) condition of constant osmotic pressure.

4.7.11. *Order of the transition and the critical point*

Below the transition there are two coexisting phases. This is a general feature of any system with a conserved order parameter. Below the transition the mean value of the concentration x is fixed at x_0 and the proportions of the two phases, of concentrations x_1 and x_2 are determined by the requirement that the mean density be equal to x_0; this is the content of the lever rule.

So the order parameter density for a system with conserved order parameter has two distinct values below the transition. Of course, this ignores the variation at the boundary between the two phases. For the binary mixture these two values would be x_1 and x_2.

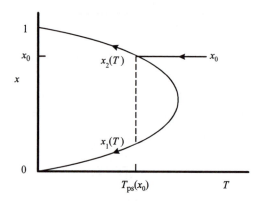

Fig. 4.46.   Order parameter values of the transition.

Above the transition the order parameter density is a constant x_0. On passing through the transition φ_1 is continuous, but φ_2 is discontinuous. It is the discontinuity that characterises the transition as being first-order. But we reiterate that since x_0 is constant the entropy is continuous, the thermal capacity is not infinite, and there is no latent heat.

Things are different, however, at $x = 1/2$. Here the order parameter density values are both continuous at the transition so this qualifies as a *second-order* transition. We also observe that the phase separation point and the spinodal point coincide so that there is no metastable region, and so no hysteresis. And finally since both the first and the second derivatives of the Helmholtz free energy (of mixing) vanish the minimum is anomalously broad so that there can be large fluctuations in the order parameter (with negligible free energy cost). We recall all these as characteristics of a second-order transition. The point at $x = 1/2$, where the transition becomes second-order is thus called the *critical point*.

We have the general rule that a first-order transition becomes second-order at a critical point. It was the assumption of a strictly regular solution that placed the critical point at $x = 1/2$. In the general case the critical point occurs at the concentration for which the transition temperature is a maximum (stationary).

4.7.12. *The critical exponent β*

The critical exponents refer to the critical point so we shall examine the binary mixture specifically in the $x_0 = 1/2$ case. It is convenient to utilise an order parameter that grows from zero at the critical point so we shall define

$$\varphi = x - 1/2.$$

Then the phase separation curve

$$T = \frac{2(1 - 2x)}{\ln[(1 - x)/x]} T_c$$

may be expressed in terms of φ as

$$\frac{T}{T_c} = \frac{2(1 - 2x)}{\ln\left(\frac{1}{2} - y\right) - \ln\left(\frac{1}{2} - y\right)}.$$

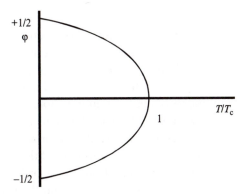

Fig. 4.47.   Critical behaviour.

We would like to expand φ in powers of $1 - T/T_c$ to obtain the order parameter critical exponent β. This is most conveniently done by expanding $1 - T/T_c$ in powers of φ and then inverting the series. We find

$$1 - \frac{T}{T_c} = \frac{4}{3}\varphi^2 + \frac{64}{45}\varphi^4 + \cdots$$

which may be inverted to yield

$$\varphi = \frac{\sqrt{3}}{2}\left(1 - \frac{T}{T_c}\right)^{1/2} - \frac{\sqrt{3}}{5}\left(1 - \frac{T}{T_c}\right)^{3/2} + \cdots .$$

In other words we have found that

$$\varphi \sim \left(1 - \frac{T}{T_c}\right)^{1/2}$$

at the critical point, the usual mean field value of $\beta = 1/2$.

Finally, we must mention an important distinction between systems with conserved and non-conserved order parameters at the critical point. The diagram above, showing the critical point may be similar in both cases, but the interpretation is different. For a non-conserved order parameter when the critical point is approached there is a *bifurcation* and the system breaks symmetry by choosing one rather than the other branch. By contrast, for a conserved order parameter *both* branches are chosen; we have coexisting fractions determined by the lever rule.

4.8.   Quantum Phase Transitions

4.8.1.   *Introduction*

In a conventional second-order phase transition there is an ordered phase existing at low temperatures. As the temperature is raised this order is destroyed by thermal fluctuations. And at the critical point the order disappears completely. The temperature of the critical point is determined by the competing requirements of energy minimisation, favouring order, and entropy maximisation, favouring disorder. Our consideration is restricted to second-order phase transitions where the "anomalous broadening" of the free energy minimum allows large excursions of the order parameter at negligible cost. These diverging fluctuations in the order parameter result in a collapse of the order and we can thus say that the transition is *driven* by the fluctuations.

Temperature is the control parameter in a conventional phase transition; as the temperature is varied the critical point is traversed. By contrast a *quantum phase transition* occurs at zero temperature. Here the control parameter is some other variable such as pressure, magnetic field, alloy composition, and the transition occurs at $T = 0$ as the control parameter is varied. At $T = 0$ there are no thermal fluctuations present and the transition from the ordered phase to the disordered is driven purely by quantum-mechanical fluctuations. Such fluctuations are a consequence of the Uncertainty Principle.

The study of quantum phase transition thus concerns the different kinds of ground state a system can have for different values of the control parameter. There will be critical exponents associated with the quantum critical point and these display the same sorts of universality as do the conventional critical exponents.

As the temperature is increased from $T = 0$, the thermal fluctuations will make an increasing contribution to the critical behaviour. In the temperature–control parameter plane there will thus be a region of ordered phase and a region of disordered phase separated by a phase boundary line. While at $T = 0$ the critical point is purely quantum, at the higher temperatures where the transition occurs the critical point will be classical. And the classical critical exponents are related rather simply to the quantum critical exponents — as we shall see.

The figure shows an example of such a transition.[17] The material is LiHoF$_4$, where the electrons in Ho are coupled by an Ising-type interaction. The control parameter is a magnetic field applied transverse to the

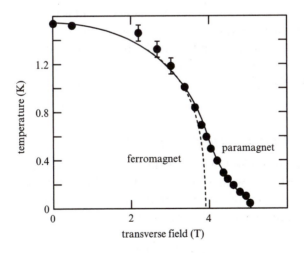

Fig. 4.48. Phase boundary in a transverse Ising system.

Ising spin direction. The dotted curve shows the result of calculation based on a mean field approximation of the Ising interaction. The solid line incorporates the effects of the nuclear hyperfine interaction as well.

4.8.2. *The transverse Ising model*

The transverse Ising model is a simple system exhibiting a quantum critical point. In this model the Ising interaction is supplemented by a magnetic field transverse to the Ising direction. Thus we write the Hamiltonian as

$$\mathcal{H} = -2\hbar J \sum_{i>j \text{ nn}} S_z^i S_z^j - \gamma \hbar B_x \sum_i S_x^i.$$

The first term is the Ising interaction and the second term represents the interaction with an applied magnetic field in the x direction, B_x.

In the absence of the external field we have the conventional Ising model. The interaction favours parallel spins and the ground state then corresponds to aligned spins either in the $+z$ or the $-z$ direction. When the transverse field is applied its effect is to induce transitions between the up and down spin states. This is because the Hamiltonian no longer commutes with the spin z component

$$[\mathcal{H}, S_z^i] \neq 0,$$

and using the Heisenberg equation of motion.

Thus the effect of the transverse magnetic field is to cause fluctuations in the z component of the spins. And these fluctuations will act to destroy the order of the Ising ground state. For high enough values of the transverse field the order will be fully destroyed and then the ground state will be qualitatively different. Thus the ground state is altered by the control parameter.

4.8.3. *Revision of mean field Ising model*

For clarity let us revise the mean field treatment of the conventional Ising model, described now in a form suitable for the addition of a transverse field. In the absence of interactions the magnetisation of an assembly of N spin $1/2$ magnetic moments μ is given by

$$M = N\mu \tanh\left(\frac{\mu B}{kT}\right)$$

where T is the temperature. Clearly the magnetisation will point in the direction of the applied magnetic field. The saturation magnetisation M_0 is given by

$$M_0 = N\mu$$

so we may write the magnetisation as

$$M = M_0 \tanh\left(\frac{M_0}{N}\frac{B}{kT}\right).$$

In the mean field approach to the Ising hamiltonian the $S_z^i S_z^j$ interaction is approximated by the average $nS_z^i\langle S_z\rangle$ where n is the number of neighbours surrounding each particle. In terms of the magnetisation this may be written as a magnetic field **b**:

$$\mathbf{b} = \lambda M_z \hat{\mathbf{z}}.$$

(Since the magnetisation is defined as the total magnetic moment, the dimensions of λ are μ_0 over volume.) Spontaneous magnetisation then occurs, in the absence of B, as the solution to

$$M_z = M_0 \tanh\left(\frac{M_0}{N}\frac{\lambda M_z}{kT}\right)$$

or

$$\frac{M_z}{M_0} = \tanh\left(\frac{M_z}{M_0}\frac{T_c}{T_0}\right)$$

where T_c, the critical temperature is given by

$$T_c = \frac{\lambda M_0^2}{Nk}.$$

The tanh equation is solved by writing it as

$$\frac{M_z}{M_0}\frac{T_c}{T} = \tanh^{-1}\left(\frac{M_z}{M_0}\right)$$

$$= \frac{1}{2}\ln\left(\frac{1+M_z/M_0}{1-M_z/M_0}\right),$$

and then

$$\frac{T}{T_c} = \frac{2M_z/M_0}{\ln\left(\dfrac{1+M_z/M_0}{1-M_z/M_0}\right)}.$$

When plotted, this gives the usual form for the variation of spontaneous magnetisation with temperature.

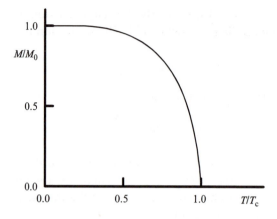

Fig. 4.49. Spontaneous magnetisation of the Ising ferromagnet.

The critical behaviour may be found from a power series expansion, since M_z is vanishingly small. The expansion is

$$\frac{T}{T_c} = 1 - \frac{1}{3}\left(\frac{M_z}{M_0}\right)^2 + \frac{4}{45}\left(\frac{M_z}{M_0}\right)^4 - \frac{44}{945}\left(\frac{M_z}{M_0}\right)^6 + \cdots$$

and this series may be inverted to yield

$$\frac{M_z}{M_0} = \sqrt{3}\left(1 - \frac{T}{T_c}\right)^{1/2} + \frac{2}{5}\sqrt{3}\left(1 - \frac{T}{T_c}\right)^{3/2} + \frac{12}{175}\sqrt{3}\left(1 - \frac{T}{T_c}\right)^{5/2} + \cdots.$$

Thus we see that the magnetisation is continuous as $T \to T_c$; the transition is second-order; the critical exponent β has the mean field value of $1/2$ and its critical amplitude is $\sqrt{3}$.

4.8.4. *Application of a transverse field*

We now have the applied magnetic field

$$\mathbf{B} = B_x \hat{\mathbf{x}}$$

to which must be added the Ising mean field

$$\mathbf{b} = \lambda M_z \hat{\mathbf{z}}.$$

Thus the total magnetic field has magnitude

$$B_{\text{tot}} = \sqrt{B_x^2 + \lambda^2 M_z^2}$$

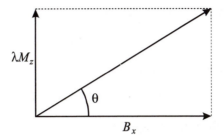

Fig. 4.50.   Resultant magnetic field.

and it points in a direction θ from the x axis, where

$$\sin \theta = \frac{\lambda M_z}{\sqrt{B_x^2 + \lambda^2 M_z^2}}.$$

Now the mean field recipe says that the magnetisation is given by

$$M = M_0 \tanh \left(\frac{M_0}{N} \frac{B_{\text{tot}}}{kT} \right)$$

and it points in the direction parallel to the field. Thus

$$M = M_0 \tanh \left(\frac{M_0}{N} \frac{\sqrt{B_x^2 + \lambda^2 M_z^2}}{kT} \right).$$

However what we are interested in is the magnetisation in the z direction. We know the direction of $\mathbf{M}$: parallel to $\mathbf{B}_{\text{tot}}$. So to find the component in the z direction we require

$$M_z = M \sin \theta,$$

or

$$M_z = \frac{\lambda M_z}{\sqrt{B_x^2 + \lambda^2 M_z^2}} M_0 \tanh \left(\frac{M_0}{N} \frac{\sqrt{B_x^2 + \lambda^2 M_z^2}}{kT} \right).$$

This may be tidied up a little, but it is helpful first, to re-express in terms of reduced variables. Since M_0 is the saturation magnetisation it follows that λM_0 is the "saturation" internal field. Let us measure the transverse field in multiples of this. That is, we define

$$b_x = B_x/\lambda M_0.$$

We also define the reduced variables

$$m_z = M_z/M_0 \quad \text{and} \quad t = T/T_c.$$

Then m_z satisfies the implicit equation

$$\sqrt{b_x^2 + m_z^2} = \tanh \frac{\sqrt{b_x^2 + m_z^2}}{t}.$$

4.8.5. *Transition temperature*

The transition from the ordered phase to disordered phase corresponds to the vanishing of M_z. In general, the temperature at which the transition occurs will be a function of the transverse field. Setting $m_z = 0$ in the above equation then gives the (b_x-dependent) critical temperature in the equation

$$b_x = \tanh \frac{b_x}{t_c(b_x)}.$$

This may be inverted to give

$$\frac{b_x}{t_c(b_x)} = \tanh^{-1} b_x$$

$$= \frac{1}{2} \ln \left(\frac{1 + b_x}{1 - b_x} \right)$$

so that

$$t_c(b_x) = \frac{2b_x}{\ln \left(\dfrac{1 + b_x}{1 - b_x} \right)}.$$

In terms of the full variables this is

$$\frac{T_c(B_x)}{T_c(B_x = 0)} = \frac{2B_x/\lambda M_0}{\ln \left(\dfrac{\lambda M_0 + B_x}{\lambda M_0 - B_x} \right)}$$

and this is plotted in Fig. 4.51 below.

This curve corresponds to the dotted line plotted with the experimental data in the introduction.

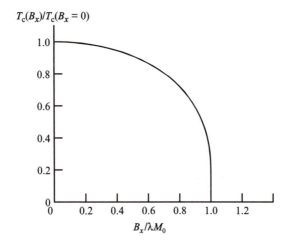

Fig. 4.51. Variation of critical temperature with transverse field.

4.8.6. *Quantum critical behaviour*

In terms of reduced variables, the equation of state for the system may be written as

$$t = \frac{2\sqrt{m_z^2 + b_x^2}}{\ln\left(\dfrac{1 + \sqrt{m_z^2 + b_x^2}}{1 - \sqrt{m_z^2 + b_x^2}}\right)}.$$

The quantum critical behaviour occurs at zero temperature. The control parameter is b_x; this is the analogue of temperature in the classical case. So to study the zero-temperature phase diagram we need the dependence of m_x on b_z.

Since we are assuming that both m_x and b_z are finite then in the limit of $t \to 0$ the logarithm must become infinite. And to achieve this, the denominator will go to zero. In other words

$$1 - \sqrt{m_z^2 + b_x^2} = 0$$

or

$$m_z^2 = 1 - b_x^2.$$

This is plotted in Fig. 4.52.

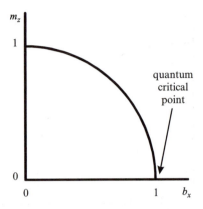

Fig. 4.52.   Zero temperature phase diagram.

To investigate the critical behaviour we want to know what is happening in the vicinity of the quantum critical point, at $b_x = 1$. We write m_z as

$$m_z = \left(1 - b_z^2\right)^{1/2}$$

and a series expansion then gives

$$m_z = \sqrt{2}(1 - b_x)^{1/2} - \frac{1}{2\sqrt{2}}(1 - b_x)^{3/2} - \frac{1}{16\sqrt{2}}(1 - b_x)^{5/2} + \cdots.$$

We see that the order parameter goes continuously to zero at the critical point so the transition is second-order. And the above expansion gives β, the order parameter critical exponent to be $1/2$. The classical and the quantum critical behaviour can be compared. We have

$$\text{classical case} \quad m_z = \sqrt{3}(1 - t)^{1/2}$$
$$\text{quantum case} \quad m_z = \sqrt{2}(1 - b_x)^{1/2}.$$

We see that the critical *exponents* are the same in the quantum and the classical case but the critical *amplitudes* are different.

4.8.7. *Dimensionality and critical exponents*

There is a general rule relating classical and quantum critical exponents and the dimensionality of the system. The scaling theory arguments outlined in Sec. 4.1.8 and in particular, the Josephson critical exponent law demonstrate the importance of the system's dimensionality to the critical exponents.

Let us write the Boltzmann factor appearing in the system's partition function as $e^{-E\beta}$ where $\beta = 1/kT$. This has a certain similarity to the quantum generator of time evolution $e^{iEt/\hbar}$ and the generator of spatial displacement $e^{ipx/\hbar}$. We can thus regard the Boltzmann factor as a generator of an (imaginary) spatial displacement. And as the temperature goes to zero the spatial displacement goes to infinity — what one understands as the thermodynamic limit. The sum or integral over Boltzmann factors in the partition covers the spatial extent of the system. And in the limit of zero temperature the Boltzmann factors involve an additional dimension to be traversed.

Since it is understood that the partition function contains all thermodynamic information about a system the above argument implies that a zero-temperature critical point in a system of n dimensions will have the same behaviour as a conventional critical point in a system of $n + 1$ dimensions. And since the mean field approximation is independent of the system's dimensionality we conclude that the mean field approximation will give the same critical exponents for the classical and the corresponding quantum critical point.

We know that mean field approaches give behaviour that, while qualitatively true, are often not in quantitative agreement with observed critical behaviour. For each system (Hamiltonian) there is a marginal spatial dimensionality d^* such that when the number of dimensions is greater than d^* the results of mean field theory are exact.[18] When the number of dimensions is less than d^* then mean field theory is quantitatively wrong; then fluctuations are significant. When the number of dimensions is equal to d^* there are only logarithmic corrections to the results of mean field theory. Thus the designation *marginal* dimensionality.

The marginal dimensionality for a dipole-coupled Ising system, such as $LiHoF_4$ described above, is 3. Thus we expect mean field theory to give, to first-order, the correct critical behaviour for the classical critical behaviour. And it will then certainly be correct for the quantum critical behaviour. In other words, for *this* system, the quantum and the classical critical behaviour should be the same: the mean field behaviour. We saw above that the mean field calculation gave the same value for the exponent β: one half.

We found that the susceptibility critical exponent γ had the value 1 for the mean field Ising ferromagnet. So we conclude that for $LiHoF_4$ both the classical and the quantum values of the exponent γ should have the value 1. Figure 4.53 below[17] shows susceptibility measurements in the vicinity of both the classical and the quantum critical points. The lines through the data both have a slope of -1 indicating $\gamma = 1$ in both cases.

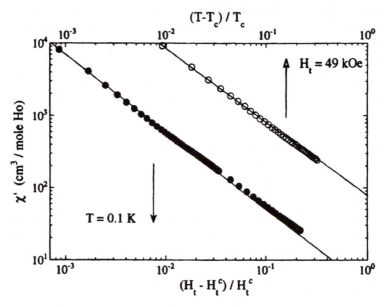

Fig. 4.53.   Quantum (below) and classical (above) critical behaviour of $LiHoF_4$.

4.9.   Retrospective

We have now completed our survey of different types of phase transitions. You will have observed similarities between systems and you will have observed differences. In this final section we shall look back over the various examples considered and we shall make some general observations and comments.

4.9.1.   *The existence of order*

Ising discovered that his model exhibited no phase transition in one dimension. In other words, in the one-dimensional Ising model displays long-range order only at $T = 0$. Onsager discovered that the Ising model did have a phase transition in two dimensions. In other words, the Ising model has a transition to an ordered state at finite temperatures in two (and higher) dimensions.

Somewhat surprisingly, Mermin and Wagner[19] argued that the Heisenberg model would not exhibit long-range order in two, but it would in three and higher dimensions. Thus the Ising model has a transition in two dimensions but the Heisenberg model does not. Why should this be? The key difference is the dimensionality of the order parameter D. For the Ising model $D = 1$ while for the Heisenberg model $D = 3$. We also note that for the spherical model ($D = \infty$), Kac reached the same conclusions as for the Heisenberg model.

The Mermin–Wagner theorem, sometimes attributed Berezinskii, Hohenberg or Peirls, states that there can be no long range order resulting from a broken continuous symmetry ($D = 2$ or above) in two or lower spatial dimension. But there can be long range order resulting from a broken discrete symmetry ($D = 1$) in a system of two spatial dimensions — the Ising model is an example of this. However in one spatial dimension there can be no long range order at all. This may be established by an energy–entropy argument similar to that in Sec. 4.4.3 above for the 1d Ising model. We may summarise as follows:

- For $d \geq 3$ you can always have an ordered phase — a broken discrete or continuous symmetry.
- For $d = 2$ you cannot break a continuous symmetry, only a discrete symmetry. So only the Ising model has a phase transition in 2d; there is no ordered phase for the Heisenberg model or the spherical model.
- For $d = 1$ you cannot break any symmetry; no ordered phase is possible in 1d above $T = 0$.

The borderline case corresponds to $d = 2$ and $D = 2$. In this case, it is possible to have *orientational* order; this is the Kosterlitz–Thouless[20] transition. Unfortunately this is outside the scope of this book. Figure 4.54 indicates the various possibilities.

Finally we mention the question (using the magnetic description) of the magnitude of the spin vector. For a spin $S = 1/2$ each moment has one of two projections along a given axis. For higher spin there will be a larger number of projections. The spin magnitude is another "dimension" which characterises systems. Of particular importance, when $S \rightarrow \infty$ the moments are classical and there are no quantum effects.

4.9.2. *Validity of mean field theory*

In Sec. 4.8.7 we learned that for each system (Hamiltonian) there is a marginal spatial dimensionality d^* such that when the number of

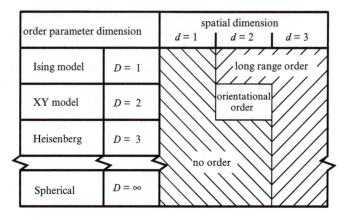

Fig. 4.54.   Dimensionality and order.

dimensions is greater than d^* the results of mean field theory are exact. We have also seen that the spatial dimension is crucial in determining whether a system will undergo a phase transition. It is therefore surprising that the spatial dimension enters nowhere in the mean field modelling of phase transitions. In fact, we saw that for a spatial dimension of four or greater the mean field description becomes exact.

It is usually the case that for short-range interactions the marginal dimensionality d^* is 4. However, when the interactions have longer range, such as dipole or Coulomb interactions, then the marginal dimensionality can be less; an example is the dipole-coupled Ising system treated in Sec. 4.8.

There is, however, a practical question when considering the validity of mean field theory. How close to the transition must one be to observe the (possible) breakdown of mean field theory? There are some systems, such as superconductors, where one needs to be impossibly close to the transition to observe the breakdown. For such systems, experimentally, they *appear* to follow mean field behaviour even though $d^* = 4$. The Ginzburg criterion specifies how close one must be; an extensive discussion of all these issues is given in the paper by Als-Nielsen and Birgenau.[18]

4.9.3. *Features of different phase transition models*

The main classification criterion we have used in discussing phase transitions has been the order of the transition — first-order or second-order.

Recall that this is essentially a generalisation/abuse of Ehrenfest's original classification; in reality we distinguish discontinuous and continuous transitions. The second classification related to whether the order parameter was conserved or non-conserved. During this chapter we have seen examples of all four possible combinations of these possibilities.

The liquid–gas transition is an example of a first-order transition in a system with a conserved order parameter. This transition has a critical point, thus one observes a second-order transition when travelling along the critical isochore. The solid–fluid transition is another example of a first-order transition in a system with a conserved order parameter. However in this case, because there is a change in symmetry between the phases, there is no critical point and the transition never becomes second-order.

The ferromagnet and the ferroelectric are both examples of transitions with non-conserved order parameters. The ferroelectric transition can be either first-order or second-order, depending on the conditions. And the changeover is at the tricritical point. The ferromagnetic transition is second-order. However, there is no reason why such a transition may not become first-order under the right circumstances. The table below summarises these and some other examples.

	First-order		Second-order	Symmetry broken
conserved order parameter	liquid–gas (but ∃ critical point)	→	liquid–gas along critical isochore	none
	binary alloy (but ∃ critical point)	→	binary alloy at critical concentration	none
	solid-fluid (no critical point)	→	- - - - - - - - -	translational invariance
non-conserved order parameter	- - - - - - - - -		ferromagnet	rotational invariance
	ferroelectric at high pressure	tricritical ← point →	ferroelectric at low pressure	inversion symmetry

Problems

4.1. Obtain an expression for the Helmholtz free energy for the Weiss model in zero external magnetic field, in terms of the magnetisation. Plot $F(M)$ for $T > T_C$, $T = T_C$ and $T < T_C$.

4.2. Show that $F = \frac{Nk}{2}\left\{(T - T_C)m^2 + \frac{T_C}{6}m^4 + \cdots\right\}$ for the Weiss model ferromagnet in the limit of small m. Explain the appearance of T_C in the m^4 term.

4.3. Show that $\mathrm{d}^2F/\mathrm{d}\varphi^2 > 0$ below T_c at the two roots $\varphi = \pm\sqrt{-F_2/2F_4}$ in the Landau model. Show that $\mathrm{d}^2F/\mathrm{d}\varphi^2 < 0$ below T_c and $\mathrm{d}^2F/\mathrm{d}\varphi^2 > 0$ above T_c at the single root $\varphi = 0$. What is the physical meaning of this?

4.4. In the Landau theory of second-order transitions calculate the behaviour of the order parameter below the critical point, $\varphi(T)$, when the *sixth*-order term in the free energy expansion is not discarded. What influence does this term have on the critical exponent β? Comment on this.

4.5. A ferroelectric has a free energy of the form

$$F = \alpha(T - T_c)P^2 + bP^4 + cP^6 + DxP^2 + Ex^2$$

where P is the electric polarisation and x represents the strain. Minimise the system with respect to x. Under what circumstances is there a first-order phase transition for this system?

4.6. Consider a one-dimensional binary alloy where the concentration of A atoms varies slowly in space: $x = x(z)$. Show that the spatial variation of x results in an additional term in the free energy per bond of $a^2\varepsilon(\mathrm{d}x/\mathrm{d}z)^2$, where a is the spacing between atoms and ε is the energy parameter defined in Sec. 4.7.3.

4.7. Show that in the vicinity of the critical point the free energy of the binary alloy may be written as

$$F_m = F_0 + 2Nk\left\{(T - T_c)\left(x - \frac{1}{2}\right)^2 + \frac{2}{3}T_c\left(x - \frac{1}{2}\right)^4 + \frac{16}{15}T_c\left(x - \frac{1}{2}\right)^6 + \cdots\right\}.$$

Discuss the Landau truncation of this expression; in particular, explain at what term the series may/should be terminated.

4.8. Plot some isotherms of the Clausius equation of state $p(V - Nb) = NkT$. How do they differ from those of an ideal gas? Does this equation of state exhibit a critical point? Explain your reasoning.

4.9. The scaling expression for the reduced free energy is given in Sec. 4.1.9 by

$$f(T, B) = A |t|^{2-\alpha} Y \left(D \frac{B}{|t|^{\Delta}} \right).$$

Show that the heat capacity is given by

$$C \sim \frac{d^2 f(t, B)}{dt^2}$$

and hence identify α as the heat capacity critical exponent.

4.10. Using the scaling expression for the reduced free energy in the previous section, show that the magnetisation is given by

$$M \sim \frac{df(t, B)}{dB}$$

and hence show that the order parameter exponent β is given by

$$\beta = 2 - \alpha - \Delta.$$

Show that the magnetic susceptibility is given by

$$\chi \sim \frac{d^2 f(t, B)}{dB^2}$$

and hence show that the susceptibility exponent γ is given by

$$\gamma = 2 - \alpha - 2\Delta.$$

4.11. Show that the Landau free energy has the scaling form of Problem 4.9 above, with $\alpha = 0$.

References

[1] C. N. Yang and T. D. Lee, *Phys. Rev.* **87** (1952) 404; T. D. Lee and C. N. Yang, *Phys. Rev.* **87** (1952) 410.

[2] K. Huang, *Statistical Mechanics*, 2nd ed. (Wiley, 1987).

[3] V. L. Ginzburg and L. D. Landau, *On the Theory of Superconductivity*, translation in *Collected Papers of L. D. Landau* (Pergamon Press, 1965), pp. 546–568.

[4] J. W. Gibbs, *Trans. Connecticut Acad. Sci.*, **3** (1875–1878) 108; *The Collected Works*, Vol. 2 (Longman and Green, NY, 1928).

[5] M. E. Fisher, *Scaling, Universality and Renormalisation Group Theory*, Lecture Notes in Physics, Vol. 186, *Critical Phenomena* (Springer-Verlag, 1983).

[6] G. F. Mazenko, *Equilibrium Statistical Mechanics* (Wiley, 2000).

[7] E. A. Guggenheim, *Thermodynamics* (North Holland, 1949); *Applications of Statistical Mechanics* (Oxford, 1966).

[8] J. Wilks, *The Properties of Liquid and Solid Helium* (Oxford, 1967).

[9] J. de Boer, *Physica* **14** (1948) 139.

[10] S. G. Brush, History of the Lenz–Ising model, *Rev. Mod. Phys.* **39** (1967) 883.

[11] L. D. Landau and E. M. Lifshitz, *Statistical Physics* (Pergamon Press, 1980).

[12] M. Plische and B. Bergersen, *Equilibrium Statistical Physics*, 2nd ed., (World Scientific, 1994).

[13] H. E. Stanley, *Introduction to Phase Transitions and Critical Phenomena* (Oxford, 1971).

[14] C. Kittel, *Introduction to Solid State Physics*, 7th ed. (Wiley, 1996).

[15] G. Burns, *Solid State Physics* (Academic Press, 1986).

[16] J. C. Slater, *Introduction to Chemical Physics* (McGraw-Hill, 1939).

[17] D. Bitko, T. F. Rosenbaum and G. Aeppli, Quantum critical behaviour for a model magnet, *Phys. Rev. Lett.* **77** (1996) 940.

[18] J. Als-Nielsen and R. J. Birgeneau, Mean field theory, the Ginzburg criterion, and marginal dimensionality of phase transitions, *Am. J. Phys.* **45** (1977) 554.

[19] D. Mermin and H. Wagner, Absence of ferromagnetism or antiferromagnetism in one- or two-dimensional isotropic Heisenberg models, *Phys. Rev. Lett.* **17** (1966) 1133–1136.

[20] J. M. Kosterlitz and D. J. Thouless, *J. Phys. C: Solid State Phys.* **6** (1973) 1181–1203.

CHAPTER FIVE

FLUCTUATIONS AND DYNAMICS

The traditional subject of *thermodynamics* is wrongly named. It deals, essentially, with equilibrium states and the relations between these equilibrium states. The existence of equilibrium states is essential to our understanding of the macroscopic world at a quantitative level. Although macroscopic systems must be described, at the microscopic level, by a very large number of variables, the equilibrium states of such systems are described in terms of only a few thermodynamic parameters. It is on this fortunate aspect of nature that the discipline of thermodynamics is based.

Classical thermodynamics relates different equilibrium states of a system. One state may have been transformed into the other by passing "quasi-statically" through a sequence of intermediate (almost) equilibrium states or the process might be completely irreversible, such as the Joule–Kelvin throttling, where all one can speak of are the initial and final equilibrium states. In the former case we can draw an "indicator diagram" showing the variation of thermodynamic variables in the process. In the latter case, all we can plot are the initial and final points. Classical thermodynamics and its microscopic underpinning of equilibrium statistical mechanics recognises the existence of initial and final states, but it can tell nothing about the *dynamics* of the transformation processes. Thus the designation *thermodynamics* is a misnomer.

Equilibrium systems have no dynamics, almost by definition. This is true "on the average", but of course, there are fluctuations. By contrast, non-equilibrium systems do have dynamics. Common experience (and indeed the Second Law of thermodynamics) tells us that a non-equilibrium system will evolve towards a state of equilibrium. The subject of non-equilibrium statistical mechanics deals with this *approach* to equilibrium.

Before making direct appeal to the microscopics of particular systems it is of interest to ask what deductions can be made from specifically macroscopic considerations. As with equilibrium "thermodynamics" the macroscopic approach can relate different properties of a system, but it cannot go much further than this. Adkins's *Equilibrium Thermodynamics*[1] discusses thermoelectricity from this perspective and the limitations of the approach are clearly discussed. Further progress in the macroscopic approach can be made through the addition of some new "laws". Traditionally Onsager's reciprocity law is used. More modern is the use of the law of minimum entropy production. Both of these are discussed in Kondepundi and Prigogine's *Moderm Thermodynamics*.[2]

The behaviour of specific systems must be treated, as in the equilibrium case, by looking at the microscopics, and applying statistical arguments. In particular, this is necessary if we want to study the time dependence of thermodynamic variables in non-equilibrium systems, and their variation as the system evolves towards equilibrium. MacDonald[3] coined the phrase "time dependent statistical mechanics" for this.

5.1. Fluctuations

5.1.1. *Probability distribution functions*

When a thermodynamic property is fixed, its conjugate variable fluctuates about a mean value. Thus we saw that when the temperature was fixed there were fluctuations in energy. In a small region of a system there will be fluctuations in all extensive variables. Here, we are regarding the small region as being a subsystem in equilibrium with a reservoir comprising the rest of the system. Thus the conditions of the equilibrium are that all intensive variables are fixed: temperature, pressure, chemical potential, etc.

To examine the fluctuations in the subsystem we must consider the appropriate thermodynamic potential, here the Gibbs free energy $G(T, p, \mu)$, since T, p, and μ are fixed. We ask the question: What is the probability of the occurrence of a fluctuation of magnitude X?

When considering an *isolated* system we saw that the Boltzmann relation $S = k \ln \Omega$ could be inverted, following Einstein, to give the probability of observing a fluctuation of magnitude X as

$$P(X) \propto e^{S(X)/k}. \tag{5.1}$$

An *open system*, that is, one having fixed values of its intensive variables is specified by its Gibbs free energy and the probability of a fluctuation of magnitude X is then

$$P(X) \propto e^{-G(X)/kT}. \tag{5.2}$$

(The probability would be given by $P(X) \propto e^{S(X)/k}$ where $S(X)$ is the total entropy of the system *plus* reservoir. It becomes $P(X) \propto e^{-G(X)/kT}$ where $G(X)$ is the Gibbs free energy of the system alone.)

In equilibrium G is a minimum.

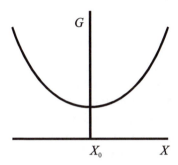

Fig. 5.1.   Equilibrium value of Gibbs free energy.

Here X could be a general extensive variable or an order parameter of a phase transition. For convenience we shall put $X_0 = 0$ so that the variable X measures the *deviation* from the mean (equilibrium) value.

For small deviations of X from its equilibrium value we may write

$$G(X) = G(0) + X \left.\frac{\partial G}{\partial X}\right|_0 + \frac{X^2}{2} \left.\frac{\partial^2 G}{\partial X^2}\right|_0 + \cdots.$$

Since the equilibrium position is a minimum, the linear term vanishes. The leading term in X is thus quadratic. To leading order in X the probability is then

$$P(X) \propto \exp - \left\{ (G(0) + G''(0)X^2/2)/kT \right\}$$

or

$$P(X) \propto e^{-G''(0)X^2/2kT}. \tag{5.3}$$

This is a Gaussian probability distribution, which has the general form

$$P(X) \propto e^{-X^2/2\langle X^2 \rangle}.$$

And from this we can identify the mean square measure of the fluctuations of X as

$$\langle X^2 \rangle = kT \left(\left. \frac{\partial^2 G}{\partial X^2} \right|_{X=X_0} \right)^{-1}. \tag{5.4}$$

We see that the *broader* the minimum in G the greater the magnitude of the fluctuations. So, in particular, at a *critical point* where we saw that $G'' \to 0$ the fluctuations become infinite. This is an important property of a critical point. We also note that the fluctuations will diverge at the *spinodal point* of a first-order transition.

5.1.2. *Mean behaviour of fluctuations*

The mean value of a thermodynamic quantity X is constant for a system in equilibrium; this is essentially the definition of equilibrium. However on the microscopic scale X will fluctuate with time, as dictated by the equations of motion of the system. The deviations of X from the mean will average to zero: equally likely to be positive or negative, but the *square* of the deviations of X from the mean will have a non-zero average. For convenience, let us redefine X by subtracting off its mean value; our newly defined X has zero mean: $\langle X \rangle = 0$. But as mentioned above, $\langle X^2 \rangle \neq 0$, and in general, certainly for even n, we will have $\langle X^n \rangle \neq 0$. These "moments" can often be calculated without too much difficulty. However they give no indication about the time dependence of the fluctuations.

[**Ensembles and averages**
It is not immediately obvious what is meant by taking an average in this case. If many copies of the system are imagined, each having the same values for the macroscopic observables then one can consider the average evaluated over this collection of copies. The imaginary collection of copies is an *ensemble* and the average is called an *ensemble average*.]

What can we say about the time variation of the fluctuations without completely solving the equations of motion for the system? Can we define a quantity that describes the mean time evolution of the variations? At a particular time t_0, we may observe that the variable X has the value a^i:

$$X(t_0) = a^i.$$

This will subsequently develop in time so that the observed mean becomes zero. We can talk of the mean time dependence of X from this value a^i

by taking a sub-ensemble from a complete ensemble — the sub-ensemble consisting of all those elements that at time t_0 have the value a^i, that is, the elements of the sub-ensemble obey

$$X^i_j(t_0) = a^i.$$

The upper index of X labels the sub-ensemble and the lower index indicates which element within the sub-ensemble. We can then define the mean regression from the initial value a^i at time t_0 by the average

$$f^i(t_0 + t) = \frac{1}{n_i} \sum_{j=1}^{n_i} X^i_j(t_0 + t)$$

where n_i is the number of elements in the ith sub-ensemble.

For sufficiently long times t, the value of $f^i(t_0 + t)$ will go to zero since there will then be no means of distinguishing that this sub-ensemble is not a representative one. Thus we expect the function $f^i(t_0 + t)$ to behave as in the figure.

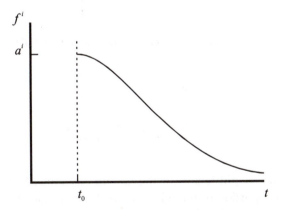

Fig. 5.2.   Mean regression of a random fluctuation from an initial value.

This function describes "on the average" how X varies if at time t_0 it had the value a^i. However since we are considering a system in equilibrium — that is, it has time translation invariance — it follows that the time evolution of $f^i(t_0 + t)$ is not peculiar to the time t_0. Whenever X is observed to take on the value a^i the subsequent mean time dependence will be given by f^i. So we may ignore t_0 and we shall consider it as zero in our future discussions.

The behaviour does, however, refer specifically to the particular value a_i. To find the mean regression of any fluctuation we may think of averaging over all the initial values a_i with appropriate weights. But this is simply an average over the complete ensemble and this, we know, must give zero. Physically we can see that this is so since there will be many positive and negative values of f^i which will average to zero.

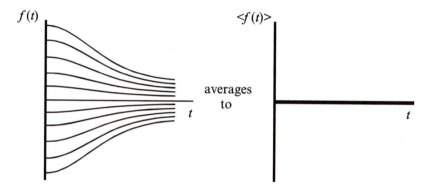

Fig. 5.3.   Average behaviour of many different initial states.

Mathematically this follows since

$$\langle f(t) \rangle = \sum_i w_i f^i(t) = \sum_i \frac{w_i}{n_i} \sum_j X_j^i(t)$$

where w_i is the weight factor for the ith sub-ensemble. Now this weight factor is the proportion of the whole ensemble that the ith sub-ensemble represents, that is n_i/N where N is the total number of elements in the ensemble: $N = \sum_i n_i$. Thus

$$\langle f(t) \rangle = \sum_i \frac{n_i}{N n_i} \sum_j X_j^i(t)$$

$$= \frac{1}{N} \sum_{i,j} X_j^i(t),$$

the common mean over the full ensemble which is zero.

Instead of the direct mean we could evaluate the mean of the squares of the regressions of the sub-ensembles. In this way, positive and negative variations will all contribute without cancelling each other. We then have

$$\langle f^2(t) \rangle = \sum_i w_i \left\{ f^i(t) \right\}^2$$

$$= \sum_i \frac{w_i}{n_i} \left\{ \sum_j X^i_j(t) \right\}^2$$

$$= \sum_i \frac{w_i}{n_i} \sum_{j,k} X^i_j(t) X^i_k(t)$$

$$= \frac{1}{N} \sum_{i,j,k} X^i_j(t) X^i_k(t)$$

but this involves cross terms between members of the ensemble. The idea of correlations between different elements of an ensemble is completely unphysical, the mathematics is becoming complicated, so we shall reject this approach.

Another possibility is to take the average of the magnitude of the $f^i(t)$, defining

$$\langle f(t) \rangle_{\text{mag}} = \sum_i w_i \left| f^i(t) \right|$$

$$= \frac{1}{N} \sum_i \left| \sum_j X^j_i(t) \right|.$$

Unfortunately there are mathematical difficulties in manipulating modulus functions, but this approach gives a good lead. In fact, this method of defining an average is equivalent to taking a mean with a weight function ε_i which is positive for positive a^i and negative for negative a^i:

$$\langle f(t) \rangle_{\text{mag}} = \frac{1}{N} \sum_i \varepsilon_i \sum_j X^i_j(x)$$

$$= \frac{1}{N} \sum_{i,j} \varepsilon_i X^i_j(x).$$

Written in this form the expression for the average behaviour of the regression seems promising. The only difficulty is with the strange weight function

$$\varepsilon_i = 1 \qquad \text{if } a^i > 0$$
$$\varepsilon_i = -1 \qquad \text{if } a^i < 0.$$

Instead of this discontinuous function ε multiplying the components, there is a much more straightforward weight function that satisfies the fundamental requirement of respecting the sign of a^i: we could use a^i themselves. In other words, in evaluating the average behaviour of the natural fluctuations we weight each element of the ensemble by its initial value. We have then

$$\frac{1}{N} \sum_{i,j} a^i X_j^i(t)$$

or, since $X_j^i(0) = X^i(0) = a^i$, it can be written as

$$\frac{1}{N} \sum_{i,j} X_j^i(0) X_j^i(t)$$

and in this form we have achieved an important progress in that the average is over the complete ensemble (with no cross terms) and can be written as

$$\langle X(0) X(t) \rangle.$$

This is our expression for the "average" regression of a fluctuation from some initial value back to the mean.

5.1.3. *The autocorrelation function*

The above expression is the mean over the ensemble where each element is weighted in proportion to its initial value. This function of the randomly varying quantity $X(t)$ is known as the autocorrelation function. We shall denote it by the symbol $G(t)$:

$$G(t) = \langle X(0) X(t) \rangle. \tag{5.5}$$

In Fig. 5.4, we show the typical behaviour for the fluctuation of a thermo-dynamic variable about its mean value.

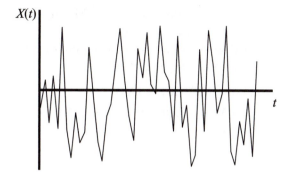

Fig. 5.4. Fluctuations in X as a function of time.

The corresponding autocorrelation function is shown below.

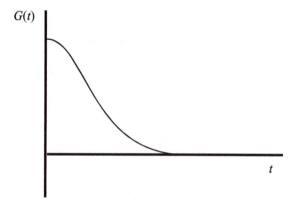

Fig. 5.5. Autocorrelation function of X.

In the sense described above, the function $G(t)$ describes the mean time variation of the fluctuations in $X(t)$. Observe the smooth behaviour of the autocorrelation function. In a sense this function has distilled the fundamental essence of the random function $X(t)$ without its mass of unimportant fine detail.

The zero time value of $G(t)$ has an immediate interpretation. From the definition of $G(t)$ we have

$$G(0) = \langle X^2 \rangle \qquad (5.6)$$

the mean square value of the fluctuating variable. For long times, as we have argued above, $G(t)$ must go to zero.

If X is not defined so that its average $\langle X \rangle$ is zero then we must modify the definition of the autocorrelation function to

$$
\begin{aligned}
G(t) &= \langle (X(0) - \langle X \rangle)(X(t) - \langle X \rangle) \rangle \\
&= \langle X(0)X(t) \rangle - \langle X(0)\langle X \rangle \rangle - \langle \langle X \rangle X(t) \rangle + \langle \langle X \rangle \langle X \rangle \rangle \\
&= \langle X(0)X(t) \rangle - \langle X \rangle^2,
\end{aligned}
$$

since

$$
\langle \langle X \rangle X(t) \rangle = \langle X \rangle \langle X(t) \rangle = \langle X \rangle^2.
$$

The time-translation invariance property, which we stated to be a property of equilibrium systems now becomes the stationarity principle:

$$
\langle X(\tau)X(t+\tau) \rangle = \langle X(0)X(t) \rangle :
$$

i.e.

$$
\text{equilibrium} \;\Rightarrow\; \text{stationarity}.
$$

And stationarity implies the time-reversal behaviour. From stationarity we have:

$$
\langle X(-\tau)X(0) \rangle = \langle X(0)X(\tau) \rangle
$$

but classically the X commute, so that

$$
\langle X(-\tau)X(0) \rangle = \langle X(\tau)X(0) \rangle
$$

or

$$
G(t) = G(-t). \tag{5.7}
$$

[A paradox

The correlation function $G(t)$ has the physical significance that a "large" fluctuation will be expected on average to die out according to $G(t)$. A large fluctuation will occur infrequently and on the "large" scale X may be assumed to decay to its equilibrium value.

We say that $G(t)$ traces the mean decay of a fluctuation. Now according to this view one would expect $G(-t)$ to tell us of the past history of the fluctuation. But we have the time-reversal rule $G(t) = G(-t)$. Our zero of time always seems to coincide with the peak value of the fluctuation — but how can this be consistent with time-translation invariance? There seems to be a paradox; the time origin is important. The resolution is that the zero of time is unique because according to the values of X at *this* time the sub-ensembles are selected. That is, the weighting factors of the averaging procedure are selected at the zero of time.]

5.1.4. *The correlation time*

The autocorrelation function, as we have seen, starts at a finite value and it decays to zero as the time argument increases. This indicates to us the average way that a fluctuation dies out. The correlation function indicates the time scale of the variations of the random variable. We quantify this time scale by introducing a *correlation time* τ_c such that for times much shorter than this $G(t)$ will hardly have changed, while times much longer than this $G(t)$ will have gone to zero.

$$G(t) \approx G(0) \quad t \ll \tau_c$$
$$G(t) \approx 0 \quad t \gg \tau_c.$$

The correlation time is indicated in the figure.

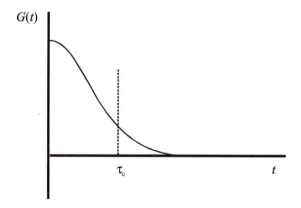

Fig. 5.6. Correlation time of an autocorrelation function.

The above statements about the correlation time do not give a recipe for its numerical evaluation; it simply describes what happens at times much shorter and much longer. It is convenient to have a precise mathematical specification and we shall see that the definition we adopt will make direct connection with future considerations.

The correlation time is a "rough measure" of the width of the correlation function. Now a rough measure of the area of the correlation function is a rough measure of its width multiplied by a rough measure of its height. A rough measure of the height of the correlation function is its initial height $G(0)$. And its area is most conveniently expressed as the integral. Thus, we are saying

$$\int_0^\infty G(t)\mathrm{d}t = \tau_c G(0)$$

or

$$\tau_c = \frac{1}{G(0)} \int_0^\infty G(t)\mathrm{d}t. \qquad (5.8)$$

This expression is taken as the definition of the correlation time.

The autocorrelation function is an important function of a random variable. We will encounter autocorrelation functions frequently in what follows. We will also see that the area under the autocorrelation function is an important quantity and because of this we will make frequent use of the correlation time as defined above.

5.2. Brownian Motion

In 1828, Robert Brown observed that tiny grains of pollen immersed in water underwent a perpetual random motion. He went on to observe this effect in a whole range of powdered substances. Brown's conclusion was that the motion was due to some "life force" in all inorganic matter. Today, the importance of his work is in the *universality* of the effect. It was Einstein in 1905,[4] who explained the origin of the motion as arising from the constant bombardment of the particles by the atoms or molecules of the fluid in which it is immersed.

The key point about Brownian motion is that it is the motion of a *macroscopic* body arising from impacts from atoms/molecules of the surrounding fluid.

The macroscopic body, which nevertheless might be quite small, may be pollen grains in water, smoke particles in air, particles of ink pigment in water, the mirror of a traditional galvanometer. The random motion of all these may be observed. For the present, we shall not consider examples such as the galvanometer mirror where there is a "restoring force" acting on the Brownian particle.

Fig. 5.7. Brownian motion of a particle in two dimensions.

5.2.1. *Kinematics of a Brownian particle*

For simplicity we shall consider motion of a Brownian particle in one dimension; the generalisation to two and three dimensions is straightforward. The location of the particle is not bounded and as time increases it will travel further and further from the origin. The square of the displacement for a typical trajectory is shown in the figure. We see that the mean square displacement increases (even though in one dimension there is always a finite probability of the particle returning to the origin).

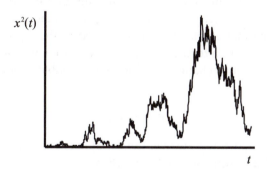

Fig. 5.8. Typical squared displacement of Brownian particle.

We shall investigate the mean square displacement of the particle. At this stage we consider the *kinematics* of the particle. That is, we will treat things at a descriptive (but quantitative) level without delving into the dynamics and the forces involved.

The distance travelled by the Brownian particle in a time t may be found by integrating up its velocity:

$$x(t) = \int_0^t v(\tau)\mathrm{d}\tau.$$

Here $v(\tau)$ is the particle's velocity at time τ.

The square of the displacement is then

$$\begin{aligned}
x^2(t) &= \left\{ \int_0^t v(\tau)\mathrm{d}\tau \right\}^2 \\
&= \int_0^t v(\tau_1)\mathrm{d}\tau_1 \int_0^t v(\tau_2)\mathrm{d}\tau_2 \\
&= \int_0^t \mathrm{d}\tau_1 \int_0^t \mathrm{d}\tau_2 v(\tau_1)v(\tau_2)
\end{aligned}$$

so the mean square displacement is

$$\langle x^2(t) \rangle = \int_0^t \mathrm{d}\tau_1 \int_0^t \mathrm{d}\tau_2 \langle v(\tau_1)v(\tau_2) \rangle.$$

We see that the mean square displacement is given in terms of the velocity autocorrelation function.

$$G_v(\tau_1 - \tau_2) = \langle v(\tau_1)v(\tau_2) \rangle.$$

Here we have used the subscript v to indicate that it is the autocorrelation function of the *velocity*. The stationarity of the random velocity (a consequence of thermal equilibrium) is indicated in the argument $(\tau_1 - \tau_2)$. This allows us to take a further step in the expression for the mean square displacement. We may change variables in the double integral to

$$\tau = \tau_1 - \tau_2$$
$$T = \tau_1 + \tau_2$$

whereupon we may integrate over the variable T. This is a non-trivial procedure, detailed in Appendix 4; the result is

$$\langle x^2(t) \rangle = 2 \int_0^t (t - \tau)G_v(\tau)\mathrm{d}\tau. \tag{5.9}$$

This is a useful expression, as we shall see. It is worthwhile to re-emphasise what has been achieved at this stage. Using only kinematics we have found an expression for the mean square of the Brownian particle in terms of the particle's velocity autocorrelation function. This also reinforced the idea that autocorrelation functions are useful quantities when considering random processes.

Pursuing the kinematical arguments further we will see that the above expression for the mean square displacement can be simplified in two limiting cases.

5.2.2. *Short time limit*

The natural time scale for the process we are considering (the only time scale we have at this stage) is the velocity correlation time, which we shall denote by τ_v. Recall we have the definition, Eq. (5.8):

$$\tau_v = \frac{1}{G_v(0)} \int_0^\infty G_v(t)\mathrm{d}t.$$

More particularly, we have the statement that

$$G_v(t) \approx G_v(0) \quad t \ll \tau_v;$$

the autocorrelation function will have changed negligibly from its initial value. So when we consider times much shorter than τ_v, we may replace $G_v(t)$ by $G_v(0)$ in the integral expression for $\langle x^2(t) \rangle$. But in this case $G_v(0)$ comes out of the integral and we have

$$\langle x^2(t) \rangle = 2G_v(0) \int_0^t (t - \tau)\mathrm{d}\tau.$$

Now the integral may be evaluated simply, giving

$$\begin{aligned}
\langle x^2(t) \rangle &= 2G_v(0) \left[t \int_0^t \mathrm{d}\tau - \int_0^t \tau \mathrm{d}\tau \right] \\
&= 2G_v(0) \left[t^2 - t^2/2 \right] \\
&= G_v(0)t^2.
\end{aligned}$$

The mean square displacement is proportional to the square of the time interval. We may write our result as

$$\langle x^2(t) \rangle = \langle v^2 \rangle t^2. \tag{5.10}$$

This indicates that the Brownian particle is moving essentially freely; at these short times there have not been sufficient atomic impacts to have any significant effect on the particle. This is referred to as the *ballistic* regime.

5.2.3. *Long time limit*

The other statement about the correlation time is

$$G_v(t) \approx 0 \quad t \gg \tau_v;$$

the autocorrelation function will have decayed to zero at long times. So when we consider times much longer than τ_v, $G_v(t)$ will be zero and in the expression for $\langle x^2(t) \rangle$ we will make negligible error by extending the upper limit of the integral to infinity

$$\langle x^2(t) \rangle = 2 \int_0^\infty (t - \tau) G_v(\tau) d\tau.$$

The integral may be rearranged as

$$\langle x^2(t) \rangle = 2t \int_0^\infty G_v(\tau) d\tau - 2 \int_0^\infty \tau G_v(\tau) d\tau.$$

The second term (independent of time) is negligible compared with the first at long times, so we conclude that in the long time limit

$$\langle x^2(t) \rangle = 2t \int_0^\infty G_v(\tau) d\tau. \tag{5.11}$$

Now we see that the mean square displacement of the Brownian particle is proportional to time (rather than the t^2 of the ballistic regime).

You should recall that a mean square displacement proportional to time is characteristic of a diffusive process. And in fact in 1d, the solution of the diffusion equation gives directly

$$\langle x^2(t) \rangle = 2Dt \tag{5.12}$$

where D is the diffusion coefficient.

Thus we conclude that in the long time limit the motion of the Brownian particle is diffusive, and its diffusion coefficient is given by

$$D = \int_0^\infty G_v(t) dt$$

or

$$D = \int_0^\infty \langle v(0)v(t) \rangle dt.$$

The diffusion coefficient is given by the area under the velocity autocorrelation function. The long time limit, when $t \gg \tau_v$, is called the diffusive regime.

Recall that the definition of the correlation time was given in terms of the area under the autocorrelation function, Eq. (5.8):

$$\tau_v = \frac{1}{G_v(0)} \int_0^\infty G_v(t)\mathrm{d}t.$$

From this, it follows that we may write the diffusion coefficient as

$$D = G_v(0)\tau_v$$

or

$$D = \langle v^2 \rangle \tau_v. \tag{5.13}$$

Again we re-emphasise that the preceeding discussion is purely kinematical. All the quantities we have considered are properties of the Brownian particle. The random atomic bombardment causes the velocity of the particle to vary randomly but we have not, as yet, considered the dynamics of the collision processes.

Although we shall not consider the dynamics of the collision processes in this section, since the system of Brownian particle plus surrounding fluid is regarded as being in thermal equilibrium, we may apply the equipartition theorem to the Brownian particle. The objection might be raised that equipartition is a classical result which becomes invalid when issues of indistinguishability and multiple occupation of states becomes important. However, we will apply equipartition specifically to the Brownian particle, not to the surrounding medium. And since the Brownian particle is a macroscopic object its behaviour may be understood purely in classical terms. Since it is in thermal equilibrium with a bath at a temperature T the equipartition theorem tells us that

$$\frac{1}{2}M\langle v^2 \rangle = \frac{1}{2}kT \tag{5.14}$$

in one dimension, where M is the mass of the Brownian particle. The mean square velocity is then

$$\langle v^2 \rangle = \frac{kT}{M}. \tag{5.15}$$

This is a *consequence* of the microscopic atomic bombardment from the surrounding fluid, but the expression is a purely thermodynamic result independent of the details of the interaction. It is sufficient that thermal equilibrium is established.

Equipartition allows us to write the diffusion coefficient of the Brownian particle as

$$D = \frac{kT}{M}\tau_v. \tag{5.16}$$

This does not mean that the diffusion coefficient is proportional to temperature since the velocity correlation time will, in general, depend on temperature. At a certain level this is still a kinematical result about the Brownian particle since τ_v is also a property of the Brownian particle. What we really want to know is how the interactions with the atoms of the surrounding medium affect the particle's motion. For this we need to consider the dynamics of the process.

5.3. Langevin's Equation

5.3.1. *Introduction*

The task of examining the *dynamics* of Brownian motion was initiated by Langevin in 1908. Langevin wrote down an equation of motion for the Brownian particle. Essentially, this was an equation of the form $F = ma$, but Langevin's important contribution was in the way he viewed the force acting on the Brownian particle. Our treatment is inspired by the papers of Uhlenbeck and Ornstein[5] and Wang and Uhlenbeck.[6]

Langevin wrote the force acting on the particle as

$$F(t) = f(t) - \frac{1}{\mu}v. \tag{5.17}$$

He regarded the force $F(t)$ acting on the particle as being made up of two contributions: a random part $f(t)$ and a systematic or friction force proportional to and opposing the particle's velocity v. The constant μ in the friction force is known as the *mobility*. This view of the forces acting is eminently sensible; we know that there will be random atomic bombardments and a body moving in a fluid is known to experience friction.

The Langevin equation is written

$$M\frac{dv(t)}{dt} = f(t) - \frac{1}{\mu}v(t) \tag{5.18}$$

and this must be solved for the velocity $v(t)$.

Considering the specific problem of Brownian motion as outlined in the previous sections, it is to be expected that by solution of the Langevin equation an expression for the velocity autocorrelation function may be found in terms of the random force $f(t)$. This will be done below.

The Langevin equation is, however, capable of much more. In particular, it will give a relation between the random force and the friction force. This is a result of considerable generality and importance since it connects in a fundamental way the random fluctuations in the system $f(t)$ and the dissipation characterised by the friction (or the mobility). This connection, in its general form, is known as the *fluctuation dissipation theorem*.

5.3.2. *Separation of forces*

It is commonly stated that it is a *hypothesis* of Langevin's approach that the force on the Brownian particle may be decomposed as the sum of a random part and a systematic friction part proportional to velocity. We shall see that this decomposition may actually be justified and understood in terms of the different centres of mass frames of the fluid and the Brownian particle.[7]

A Brownian particle at rest in the centre of mass frame of the fluid medium suffers bombardments from the atoms of the fluid. These bombardments will result in a random force. On average there will be as many impacts in each direction so the average of the force will be zero.

Now consider the particle moving with respect to the centre of mass frame of the fluid. Then the impacts from the front will be at a greater relative velocity and the impacts from the rear will be at a lesser relative velocity. This will result in a mean force acting on the particle in opposition to its motion. We can see this from a simple model. Let us consider two impacts, one from the rear and one from the front, where the atoms are moving with velocities $+v$ and $-v$. with respect to the fluid centre of mass.

Fig. 5.9. Bombardment of Brownian particle.

The impact from the atom to the left transfers momentum $m\Delta v$ to the Brownian particle. Assuming the atom mass m is very much less than that of the Brownian particle M, its velocity will be reversed. Its change of velocity is then twice the relative velocity, $2(v - V)$, so the momentum transferred is then

$$\Delta p_{\text{left}} = 2m(v - V).$$

In the impact from the right the change of velocity of the atom will similarly be twice the relative velocity, in this case $2(v + V)$. So the momentum transferred from this impact is

$$\Delta p_{\text{right}} = 2m(v + V).$$

The net momentum transfer is the difference between these

$$\Delta p = -4mV.$$

Then if there are n impacts per unit time the net force on the Brownian particle will be

$$-4nmV.$$

On average, in the centre of mass frame of the fluid, there will be equal impacts from the left and right, leading to an average force of the above form.

It is of interest to observe that we have *derived* a force of friction proportional to velocity. This is in contrast to elementary discussions which state that the friction force is *assumed* to be proportional to velocity. The hidden assumption in our treatment is that the motion of the Brownian particle does not disturb the atomic motions of the fluid. This should be true if the velocity is small or the fluid is not too dense. This is, essentially, a "linear" approximation since if the motion does affect the fluid then that will feed back as an additional force on the particle, changing its motion, which will feed back to the fluid and ... round and round.

We have seen that Langevin's decomposition of the forces acting on the Brownian particle may be understood in terms of the different centres of mass frames of the fluid and the particle. The random force $f(t)$, whose mean value is zero, is supplemented by a mean force proportional to and opposing the velocity of the particle. Thus, we have justified writing the force as

$$F(t) = f(t) - \frac{1}{\mu}v$$

and we see that the mobility μ should be related to $f(t)$ in a fundamental way. This relation will follow from our consideration of the solution of the Langevin equation.

5.3.3. *The Langevin equation*

We write the Langevin equation as

$$M\frac{dv(t)}{dt} + \frac{1}{\mu}v(t) = f(t), \tag{5.19}$$

which emphasises its structure as an inhomogeneous linear first-order ordinary differential equation with source $f(t)$. It is convenient to make a simplification by the substitutions

$$A(t) = \frac{f(t)}{M}$$

$$\gamma = \frac{1}{M\mu}.$$

Then the Langevin equation becomes

$$\frac{dv(t)}{dt} + \gamma v(t) = A(t).$$

This has solution

$$v(t) = v(0)e^{-\gamma t} + \int_0^t e^{\gamma(u-t)}A(u)du.$$

The first term represents the transient part of the solution: that which depends on the initial conditions and which arises from the solution to the corresponding homogeneous equation. This is the complementary function. The second term represents the steady state response to the source "force" $A(t)$. This is the particular integral and this part persists when all memory of the initial condition has gone.

It is conventional to enunciate properties of the (scaled) random force $A(t)$.[5] These are listed as

1. $\langle A(t) \rangle = 0$. This follows, in our treatment, from the considerations of the centre of mass frame of the fluid.
2. $\langle A(t_1)A(t_2) \rangle = 0$ unless t_1 is "almost identical with" t_2. We understand this to mean that the correlation time of the random force is short.
3. $\langle A^2(t) \rangle$ has some definite value (independent of t).

We may develop property 2 by approximating

$$\langle A(t_1)A(t_2) \rangle = A^2\delta(t_1 - t_2). \tag{5.20}$$

If we integrate this we obtain

$$A^2 = \int_{-\infty}^{\infty} \langle A(0)A(t) \rangle dt \tag{5.21}$$

so that A^2 is the area under the (scaled) random force correlation function.

As a simple application of the above results we can examine the mean value of $v(t)$. We find for a given initial condition

$$\langle v(t) \rangle = v(0)e^{-\gamma t} \tag{5.22}$$

since by property 1 $\langle A(t) \rangle = 0$. This tends to zero as time proceeds and memory of the initial condition fades.

5.3.4. *Mean square velocity and equipartition*

By similar arguments, we can now examine the mean square velocity. A key result then follows when we exploit the equipartition theorem to relate the equilibrium mean square velocity of the Brownian particle to the temperature of its surrounding medium. The expression for the mean square velocity is

$$\langle v^2(t) \rangle = v^2(0)e^{-2\gamma t} + 2e^{-2\gamma t} \int_0^t e^{\gamma u} \langle v(0)A(u) \rangle du$$
$$+ e^{-2\gamma t} \int_0^t du \int_0^t dw e^{\gamma(u+w)} \langle A(u)A(w) \rangle.$$

The first term is the transient response which dies away at long times; it is of no interest. The second term vanishes since there is no correlation between $v(0)$ and $A(t)$. The third term is of interest since it describes the equilibrium state of the particle, independent of the initial conditions. In this term, we make use of property 2 and approximate the force autocorrelation function by the delta function expression, Eq. (5.20):

$$\langle A(t_1)A(t_2) \rangle = A^2 \delta(t_1 - t_2).$$

This forces $w = u$ when the integral over w is performed. Thus, we obtain at long times

$$\langle v^2(t) \rangle = A^2 e^{-2\gamma t} \int_0^t du e^{2\gamma u}$$

$$= \frac{A^2 e^{-2\gamma t}}{2\gamma}(e^{2\gamma t} - 1)$$

$$= \frac{A^2}{2\gamma}(1 - e^{-2\gamma t}). \tag{5.23}$$

And in the long time limit this takes on the time-independent value

$$\langle v^2 \rangle = \frac{A^2}{2\gamma}.$$

The importance of this expression becomes apparent when we exploit the equipartition theorem. This tells us, as we have seen,

$$\langle v^2 \rangle = \frac{kT}{M} \tag{5.24}$$

so that

$$\gamma = \frac{M}{2kT} A^2 \tag{5.25}$$

which provides us with a relation between the mobility (contained in γ) and the random force (contained in A). From the definition of γ, that for A^2 and that for $A(t)$, we can express the mobility as

$$\frac{1}{\mu} = \frac{1}{2kT} \int_{-\infty}^{\infty} \langle f(0)f(t) \rangle \, \mathrm{d}t. \tag{5.26}$$

This expression achieves the objective of relating the two forces in the Langevin equation, the mobility or friction force and the random force of atomic bombardment. The structure of this expression is that the systematic/dissipative force is expressed in terms of the autocorrelation function of the random/fluctuation force. This is a very general result, called the *fluctuation dissipation theorem*. The kT factor that appears in the relation between the macroscopic and the microscopic force is, recall, a consequence of equipartition.

5.3.5. *Velocity autocorrelation function*

In our kinematical analysis of Brownian motion we saw that the motion of the Brownian particle was conveniently expressed in terms of the velocity autocorrelation function. The calculation of this is only slightly more complicated than that of the mean square velocity. We have

$$\langle v(t)v(t+\tau) \rangle = v^2(0)e^{-\gamma(2t+\tau)} + e^{-\gamma(2t+\tau)} \int_0^t \mathrm{d}u \int_0^{t+\tau} \mathrm{d}w \, e^{\gamma(u+w)} \langle A(u)A(w) \rangle$$

where the cross term vanishes, as above. And as discussed above, the first term is of no interest since at long times t the memory of the initial state is lost. The steady-state behaviour is contained in the remaining term. To proceed, we use property 2 of the force autocorrelation function and the delta function approximation. This forces $w = u$ when the integral over w is performed. The calculation is identical to that for the mean square velocity, except for the additional $e^{-\gamma\tau}$ prefactor

$$\langle v(t)v(t+\tau)\rangle = \frac{A^2}{2\gamma}e^{-\gamma\tau}$$

or

$$G_v(t) = \langle v^2(0)\rangle e^{-\gamma t}. \tag{5.27}$$

Thus, we conclude that the correlation time for the velocity autocorrelation function is simply the damping time associated with the friction force

$$\tau_v = \gamma^{-1}.$$

We saw that the diffusion coefficient of the Brownian particle was given in terms of τ_v by

$$D = \frac{kT}{M}\tau_v,$$

which we can now re-express as

$$D = \frac{kT}{M\gamma}$$

or

$$D = \mu kT. \tag{5.28}$$

This connection between the diffusion coefficient and the mobility is known as the *Einstein relation*. That is fine; it is purely kinematical and descriptive. But the real advance is that the fluctuation dissipation theorem of the previous section allows us to express this in terms of the fluctuating microscopic forces. That is the true content of the Einstein relation.

[A paradox

Thermal equilibrium is intimately connected with time translation invariance. And on general grounds one can argue that correlation functions such as that for the velocity must then decay exponentially (in any time interval of a given duration the function decreases by the same factor). Our calculation above for $G_v(t)$ conforms to this. This general result is known as Doob's theorem.[8] However, we have a problem. Such behaviour is incompatible with microscopic reversibility which implies that $G_v(t) = G_v(-t)$, so that assuming $G_v(t)$ is analytic at the origin, all its odd derivatives would have to vanish. The paradox is resolved, however, when we recall we used the delta function approximation for the force autocorrelation time. This causes bad behaviour at $t = 0$. For a smoothed $G_f(t)$ we would find that $G_v(t)$ was exponential over most of its range, but for very short times it would flatten off. Doob's theorem would not apply at very short times; it is a thermodynamic/hydrodynamic result so that is no problem.]

5.3.6. *Electrical analogue of the Langevin equation*

The Langevin equation, Eq. (5.19):

$$M\frac{dv(t)}{dt} + \frac{1}{\mu}v(t) = f(t)$$

describes the velocity $v(t)$ of the Brownian particle of mass M in terms of the mobility μ (inverse friction) and the random force $f(t)$. However the real achievement of the approach was in relating the mobility to the random force, Eq. (5.26). In this section we shall explore an electrical analogue of this.

Imagine an electrical circuit comprising an inductor of inductance L and a resistor of resistance R. The current I flowing in the circuit results from the motion of a very large number of electrons. The voltage V across the circuit is given by

$$L\frac{dI(t)}{dt} + RI(t) = V(t). \tag{5.29}$$

The random motion of the electrons will result in a randomly fluctuating voltage $V(t)$. Then the analogy between the above two equations is quite clear. We can follow the analogy through the arguments of the previous sections. In particular, the analogue of the fluctuation–dissipation result, Eq. (5.26) gives

$$R = \frac{1}{2kT}\int_{-\infty}^{\infty}\langle V(0)V(t)\rangle dt. \tag{5.30}$$

This shows how the resistance (dissipation) is related to the fluctuations of voltage. This will be explored further in Sec. 5.5.2 where we shall consider Nyquist's theorem.

In Problem 5.9 you will examine a different electrical analogue of the Langein equation which results in relating resistance to the current fluctuations thus:

$$\frac{1}{R} = \frac{1}{2kT}\int_{-\infty}^{\infty}\langle I(0)I(t)\rangle dt. \tag{5.31}$$

5.4.   Linear Response — Phenomenology

5.4.1. *Definitions*

In this section we turn to the question of the response of a system to an externally applied disturbance. This is a non-equilibrium problem and so it is outside the area of applicability of equilibrium statistical mechanics. A special case of this question is the way a system evolves to its equilibrium state from an initial non-equilibrium configuration. The general case involves the introduction of the concept of the dynamic susceptibility; for magnetic systems this is a generalisation of the magnetic susceptibility discussed in Chapter 2.

We consider an arbitrary system, to which we apply a generalised force B. And we observe the response M to this force. In the magnetic case, B could be an applied magnetic field and M could be the magnetisation response. But the discussion will remain more general than this.

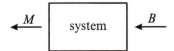

Fig. 5.10.   Excitation B and its response M.

The response depends on the force; M is a function of B. Thus we write

$$M = f(B).$$

We shall now assume that our system obeys the following three rules:

(a) Linearity
If M_1 is the response to the force B_1 and M_2 is the response to the force B_2 then the response to the force $\alpha B_1 + \beta B_2$ is given by $\alpha M_1 + \beta M_2$. In other words, the response function f obeys the rule

$$f(\alpha B_1 + \beta B_2) = \alpha f(B_1) + \beta f(B_2). \tag{5.32}$$

Physically we would expect the response to be linear for sufficiently small excitation B.

If we now consider time variation, then we expect that M at a given time will depend on B at other times. The condition of linearity means that M at a given time will depend linearly on B at other times. Thus $M(t)$ will be a linear functional of $B(\tau)$:

$$M(t) = \int X(t, \tau)B(\tau)d\tau. \tag{5.33}$$

The function X is known as the time domain dynamical susceptibility or response function.

(b) Stationarity
If the time variation of $B(t)$ is shifted by an amount t' then the response $M(t)$ will be shifted by the same amount t'. This is the physical requirement of time translation invariance. If we apply this time shift to Eq. (5.33) we find that time translation invariance requires that

$$X(t + t', \tau + t') = X(t, \tau).$$

This indicates that the response function X depends only on the time difference $t - \tau$. And then Eq. (5.33) can be written as

$$M(t) = \int X(t - \tau)B(\tau)d\tau. \tag{5.34}$$

In many respects the response function $X(t - \tau)$ describes the *memory* of the system. $M(t)$ will depend strongly on the excitation $B(\tau)$ when t is close to τ, and less so as t becomes more distant from τ.

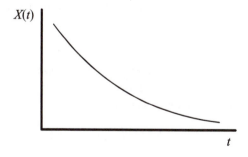

Fig. 5.11.   General form of the response function.

(c) Causality
It is a fundamental principle of Physics that effect cannot precede cause. Thus in Eq. (5.34) we see that this requires that $X(t)$ is zero for negative t:

$$X(t) = 0 \quad t < 0. \tag{5.35}$$

This requirement also means that the domain of integration in Eq. (5.34) may be taken to be $-\infty < \tau < t$.

$$M(t) = \int_{-\infty}^{t} X(t - \tau)B(\tau)d\tau. \tag{5.36}$$

This is the general expression for the linear response of M to the excitation B.

5.4.2. *Response to a sinusoidal excitation*

Many experimental techniques are based upon observing the response of a system to a sinusoidal excitation. This is the method of *spectroscopy*.

Since the system is linear, we can consider the response to the unphysical complex exponential excitation

$$B(t) = be^{-i\omega t} = b\cos(\omega t) - ib\sin(\omega t). \tag{5.37}$$

Linearity, Eq. (5.32), means that the mathematical response will be a complex $M(t)$, the real part being the response to $b\cos\omega t$ and the imaginary part being the response to $-b\sin\omega t$. The complex representation proves particularly convenient since it obviates the need to keep track of the sines and cosines. Note, however, that this simplification cannot be used when the response becomes non-linear.

The response to the complex sinusoidal excitation, Eq. (5.37) will be

$$M(t) = b\int_{-\infty}^{\infty} X(t - \tau)e^{-i\omega \tau}d\tau.$$

Here, we have expressed the integral with its upper limit as infinity; we account for causality through the requirement of Eq. (5.35). By a change of variable this result may be expressed as

$$M(t) = be^{-i\omega t}\int_{-\infty}^{\infty} X(\tau)e^{i\omega \tau}d\tau$$

$$= be^{-i\omega t}\chi(\omega). \tag{5.38}$$

This is telling us that the response to the monochromatic excitation $be^{-i\omega t}$ is the monochromatic response $\chi(\omega)be^{-i\omega t}$. Here $\chi(\omega)$ is the (complex) frequency domain dynamical susceptibility,

$$\chi(\omega) = \int_{-\infty}^{\infty} X(t)e^{i\omega t}dt. \tag{5.39}$$

It is simply the Fourier transform of the time domain dynamical susceptibility. The fact that $\chi(\omega)$ is complex is a reflection of the phase shift between
excitation and response.

As argued above, linearity means that we can separate the real and
imaginary parts. Thus

$$\Re\left\{\chi(\omega)be^{-i\omega t}\right\} \text{ is the response to } \Re\left\{be^{-i\omega t}\right\}$$

and similarly for the imaginary parts. If we write $\chi(\omega)$ in terms of its real
and imaginary parts:

$$\chi(\omega) = \chi'(\omega) + i\chi''(\omega)$$

then we see that

$$b(\chi'(\omega)\cos\omega t + \chi''(\omega)\sin\omega t) \text{ is the response to } b\cos\omega t. \tag{5.40}$$

Thus $\chi'(\omega)$ gives the in-phase component of the response and $\chi''(\omega)$ gives
the quadrature component. The phase shift is thus given by the phase
angle of $\chi(\omega)$.

5.4.3. *Fourier representation*

Any excitation $B(t)$ can be made up from a superposition of sinusoids:

$$B(t) = \frac{1}{2\pi}\int_{-\infty}^{\infty} b(\omega)e^{-i\omega t}\,d\omega.$$

Now the response to $b(\omega)e^{-i\omega t}$ is given by Eq. (5.38). Let us call this
$m(\omega)e^{-i\omega t}$. Then

$$m(\omega)e^{i\omega t} = \chi(\omega)b(\omega)e^{i\omega t}$$

so that on integrating over ω we obtain

$$\int_{-\infty}^{\infty} m(\omega)e^{i\omega t}\,d\omega = \int_{-\infty}^{\infty} \chi(\omega)b(\omega)e^{i\omega t}\,d\omega$$

where the left-hand side of this equation is the Fourier transform of $M(t)$,
that is,

$$M(t) = \frac{1}{2\pi}\int_{-\infty}^{\infty} m(\omega)e^{-i\omega t}\,d\omega.$$

Thus the equation

$$m(\omega) = \chi(\omega)b(\omega)$$

expresses the relation between the Fourier transform of the excitation and the response.

$$m(\omega) = \int_{-\infty}^{\infty} M(t)e^{i\omega t}\,dt$$

$$M(t) = \frac{1}{2\pi}\int_{-\infty}^{\infty} m(\omega)e^{-i\omega t}\,d\omega. \tag{5.41}$$

The *same* Fourier transform relations hold also for $b(\omega)$, $B(t)$ and $\chi(\omega)$, $X(t)$.

5.4.4. *Response to a step excitation*

Let us consider the response to a step excitation. We imagine a constant excitation to have been applied back into the past and we remove this at time $t = 0$. Clearly for negative times the response $M(t)$ will be a constant. And the interest then focuses on how $M(t)$ relaxes to its new equilibrium value when B drops to zero.

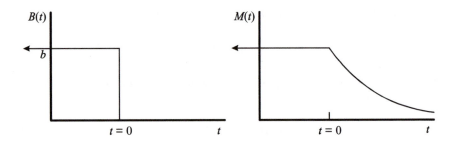

Fig. 5.12.   Step excitation and its response.

Here, the excitation is specified to be

$$\begin{aligned} B(\tau) &= b \quad \tau < 0 \\ &= 0 \quad \tau > 0 \end{aligned}$$

or

$$B(\tau) = b\,\theta_-(\tau)$$

where $\theta_-(\tau)$ is the unit step (off) function. Then the response, according to Eq. (5.34), is given by

$$M_{\text{step}}(t) = b \int_{-\infty}^{0} X(t - \tau) \mathrm{d}\tau$$

or, upon change of variables

$$M_{\text{step}}(t) = b \int_{t}^{\infty} X(\tau) \mathrm{d}\tau. \tag{5.42}$$

Of course, this holds only for times $t > 0$.

5.4.5. *Response to a delta function excitation*

Let us now consider the response to a delta function, or spike, excitation. Now $B(\tau)$ is specified as

$$B(\tau) = b\delta(\tau)$$

where $\delta(\tau)$ is the Dirac delta function.

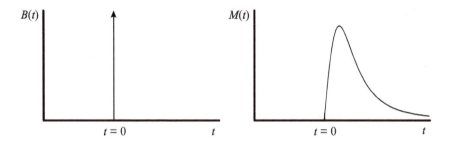

Fig. 5.13.   Delta function excitation and its response.

Then the response, according to Eq. (5.34) is given by

$$M_\delta(t) = b \int_{-\infty}^{\infty} X(t - \tau)\delta(\tau) \mathrm{d}\tau$$

and, upon using the properties of the delta function, this becomes

$$M_\delta(t) = bX(t). \tag{5.43}$$

This is just the response function $X(t)$ introduced above. In other words $X(t)$ may be regarded as the response to a delta function of unit strength.

The delta function is minus the derivative of the unit step function. The linearity condition relates the responses to these excitations. And indeed directly from Eqs. (5.42) and (5.43) we see that

$$M_\delta(t) = -\frac{\mathrm{d}}{\mathrm{d}t} M_{\text{step}}(t).$$

It will prove useful to define the function $\Phi(t)$ as the response to a unit step excitation just as $X(t)$ is the response to a delta function excitation. Then

$$X(t) = -\frac{\mathrm{d}}{\mathrm{d}t} \Phi(t). \qquad (5.44)$$

5.4.6. *Consequence of the reality of X(t)*

In this section we shall revisit the use of complex quantities in linear response questions. The response to a real time-dependent excitation $B(t)$ is a real quantity $M(t)$. And these are related by the integral expression, Eq. (5.36). Thus, we conclude that the kernel $X(t)$ must be a real function. However its Fourier transform, $\chi(\omega)$, the dynamical susceptibility, can be complex. We shall see that the reality of the time function has important consequences for the frequency function.

In Eq. (5.39):

$$\chi(\omega) = \int_{-\infty}^{\infty} X(t)e^{i\omega t}\,\mathrm{d}t$$

let us take the complex conjugate. This gives

$$\chi^*(\omega) = \int_{-\infty}^{\infty} X(t)e^{-i\omega t}\,\mathrm{d}t$$

since the real $X(t)$ remains unchanged. And if we now replace ω by $-\omega$ we have

$$\chi^*(-\omega) = \int_{-\infty}^{\infty} X(t)e^{i\omega t}\,\mathrm{d}t = \chi(\omega)$$

so we conclude that

$$\chi^*(-\omega) = \chi(\omega). \qquad (5.45)$$

This is the consequence of the reality of $X(t)$.

In terms of the real and imaginary parts of the susceptibility this gives the pair of relations

$$\chi'(\omega) = \chi'(-\omega)$$
$$\chi''(\omega) = -\chi''(-\omega).$$

(5.46)

Thus $\chi'(\omega)$ is an even function and $\chi''(\omega)$ is an odd function. This means that $\chi''(0)$ must be zero; obviously there cannot be a quadrature response at zero frequency, the static case.

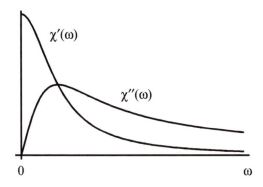

Fig. 5.14. Real and imaginary part of the dynamical susceptibility.

5.4.7. *Consequence of causality*

The requirement of causality was expressed in Eq. (5.35), that $X(t)$ is zero for negative times. We can express this condition by

$$X(t) = \theta_+(t)X(t)$$

(5.47)

where $\theta_+(t)$ is the unit step (on) function. We also have the result

$$\theta_-(t)X(t) = 0.$$

(5.48)

These equations put an important restriction on the dynamic susceptibility $\chi(\omega)$. This is most easily seen from the convolution theorem for Fourier transforms:

$$\int_{-\infty}^{\infty} F(t)\,G(t)e^{i\omega t}\,dt = \frac{1}{2\pi}\int_{-\infty}^{\infty} f(\omega')\,g(\omega-\omega')d\omega'$$

(5.49)

where $F(t)$, $f(\omega)$ and $G(t)$, $g(\omega)$ are Fourier transform pairs, as defined by Eq. (5.41).

Now the Fourier transform of the two step functions is

$$\left.\begin{array}{l} \mathcal{F}\{\theta_+(t)\} = \vartheta_+(\omega) = \dfrac{i}{\omega + i\varepsilon} \\[3mm] \mathcal{F}\{\theta_-(t)\} = \vartheta_-(\omega) = \dfrac{-i}{\omega - i\varepsilon} \end{array}\right\} \quad \varepsilon \to 0_+.$$

So the Fourier transform of Eq. (5.47) is, using Eq. (5.49):

$$\chi(\omega) = \frac{i}{2\pi} \int_{-\infty}^{\infty} \frac{\chi(\omega')d\omega'}{\omega - \omega' + i\varepsilon}$$

and from Eq. (5.48)

$$\frac{i}{2\pi} \int_{-\infty}^{\infty} \frac{\chi(\omega')d\omega'}{\omega - \omega' - i\varepsilon} = 0.$$

We define the *principal part* of the integral along the real line by

$$\mathcal{P} \int_{-\infty}^{\infty} \frac{f(\omega')d\omega'}{\omega - \omega'} = \lim_{\delta \to 0} \left\{ \int_{-\infty}^{\omega - \delta} \frac{f(\omega')d\omega'}{\omega - \omega'} + \int_{\omega + \delta}^{\infty} \frac{f(\omega')d\omega'}{\omega - \omega'} \right\}$$

and we shall accept this as equal to

$$\frac{1}{2} \left\{ \int_{-\infty}^{\infty} \frac{f(\omega')d\omega'}{\omega - \omega' + i\varepsilon} + \int_{-\infty}^{\infty} \frac{f(\omega')d\omega'}{\omega - \omega' - i\varepsilon} \right\}.$$

Then from Eqs. (A37) and (A38) we obtain

$$\chi(\omega) = \frac{i}{\pi} \mathcal{P} \int_{-\infty}^{\infty} \frac{\chi(\omega')\, d\omega'}{\omega - \omega'}.$$

The appearance of i in this equation means that if we separate the real and imaginary parts, χ' is given in terms of χ'' and *vice versa*.

$$\begin{aligned} \chi'(\omega) &= -\frac{1}{\pi} \mathcal{P} \int_{-\infty}^{\infty} \frac{\chi''(\omega')d\omega'}{\omega - \omega'} \\[3mm] \chi''(\omega) &= \frac{1}{\pi} \mathcal{P} \int_{-\infty}^{\infty} \frac{\chi'(\omega')d\omega'}{\omega - \omega'}. \end{aligned} \tag{5.50}$$

These are known as the Kramers–Kronig relations; they are purely a consequence of causality. They tell us that the real and imaginary parts of the susceptibility are not independent; the real part may be found from knowledge of the imaginary part *at all frequencies* and *vice versa*.

5.4.8. *Energy considerations*

If we regard the excitation $B(t)$ as a force and the response $M(t)$ as a displacement, then the element of work done is given by

$$dW = B\,dM. \tag{5.51}$$

And the power dissipated, the rate of doing work is then

$$\frac{dW}{dt} = B\frac{dM}{dt}.$$

For a periodic excitation we can find the mean power dissipated by averaging this expression over a cycle.

$$P = \left\langle B\frac{dM}{dt}\right\rangle.$$

From Eq. (5.40) we see that the response to an excitation $B(t) = b\cos\omega t$ is

$$M(t) = b\big(\chi'(\omega)\cos\omega t + \chi''(\omega)\sin\omega t\big).$$

Then the power dissipated is then

$$P = \omega b^2\big(-\chi'(\omega)\langle\cos\omega t\,\sin\omega t\rangle + \chi''(\omega)\langle\cos^2\omega t\rangle\big).$$

The cos sin average is zero and the $\cos^2$ average is $1/2$. Thus we find

$$P = \frac{1}{2}b^2\omega\chi''(\omega);$$

the power dissipated depends solely on the imaginary part of the dynamical susceptibility. In other words, dissipation occurs in the response that is 90° out of phase with the excitation. Since the energy dissipation must be positive, a requirement of the Second Law of thermodynamics, this puts a further general restriction on the dynamical response: $\omega\chi''(\omega)$ must be positive at all frequencies.

There is, of course, a certain arbitrariness in what we choose to call a generalised force or excitation and in what we call a generalised displacement or response. However we shall adopt Eq. (5.51) as a defining relation so that, in general, the excitation B and the response M are related by

$$B = \frac{\partial\,\text{energy}}{\partial M}. \tag{5.52}$$

Thus they are the regular extensive–intensive conjugate pairs of thermodynamics.

We can use this result to examine the dimensions of the generalised susceptibility. The defining equation for the dynamical response function gives us the result

$$[M] = [\chi][B]$$

and the energy expression above tells us that

$$[B] = [energy]/[M].$$

By eliminating $[B]$ we then find that

$$[\chi] = [M^2]/[energy];$$

the dynamical susceptibility thus has the dimensions of the square of the response divided by energy. In a similar way, we see that the dimensions of the dynamical response function are

$$[X] = [M^2]/[energy][time]$$

and the dimensions of the step response function are

$$[\Phi] = [M^2]/[energy].$$

5.4.9.   *Static susceptibility*

The static susceptibility is the response to a constant excitation. This may be regarded as the zero-frequency limit of the frequency-dependent susceptibility $\chi(\omega)$. We denote the static susceptibility by χ_0. Thus

$$\chi_0 = \chi(\omega \to 0). \tag{5.53}$$

We can relate the static susceptibility to the time domain response function through the Fourier transform relation, Eq. (5.39), taking the zero-frequency limit. Then, we find

$$\chi_0 = \int_0^\infty X(t)dt; \tag{5.54}$$

the static susceptibility is the area under the time response function. We also note that the static susceptibility is equal to the zero-time value of the step response function, defined in Sec. 5.4.5.

$$\Phi(0) = \chi_0. \tag{5.55}$$

This may be seen from Eq. (5.44).

We know that the imaginary part of χ vanishes at zero frequency. So the static susceptibility may be expressed as the zero-time value of the real part of the susceptibility:

$$\chi_0 = \chi'(\omega \to 0). \tag{5.56}$$

Now the zero-frequency value of χ' may be found from the Kramers–Kronig relation:

$$\chi_0 = \frac{1}{\pi} \int_{-\infty}^{\infty} \frac{\chi''(\omega)}{\omega}\, d\omega. \tag{5.57}$$

This is the basis of resonance methods used to measure very small static susceptibilities in terms of power absorption. Observe that there is no need to take the principal part of the integral as χ'' vanishes (sufficiently fast) at $\omega = 0$, so that there is no pole in the integral.

We finish this section by making a connection between the generalised susceptibility discussed here and the magnetic susceptibility treated in Chapter 2. The magnetic susceptibility is defined, in the SI system of units, as the ratio of the magnetisation per unit volume to the applied magnetic H field. It is not our place here to discuss the rights and wrongs of this definition. But we note two points that follow. Firstly, since we are using magnetisation *per unit volume*, this is an intensive quantity. The magnetic field is an intensive quantity and this means that the magnetic susceptibility is *intensive*. Secondly, since the magnetisation per unit volume and the H field have the same dimensions, this means that the magnetic susceptibility is dimensionless. These are the consequences of the SI definition of electromagnetic units. Now let us look at the generalised susceptibility of this chapter, in the magnetic case. The response here is the total magnetisation — an extensive quantity. The magnetic field is intensive so this means that the generalised susceptibility is *extensive*. And here the appropriate magnetic field is the B field, to ensure the product of the excitation and the response has the dimensions of energy. We then conclude that the electromagnetic susceptibility and the generalised susceptibility are related by

$$\chi_{\text{gen}} = \frac{V}{\mu_0} \chi_{\text{em}}. \tag{5.58}$$

In Chapter 2 we found the Curie law static magnetic susceptibility of an assembly of N magnetic moments μ. We can now recast the results expressed in Eqs. (2.12) and (2.13), as the generalised static susceptibility of this chapter, using Eq. (5.58). The result is

$$\chi_0 = \frac{N}{kT}\mu^2 = \frac{1}{kT}\langle M^2 \rangle. \tag{5.59}$$

5.4.10. *Relaxation time approximation*

The memory of a system to a disturbance is most directly specified in the step response function, $\Phi(t)$, schematically displayed in Fig. 5.12. This is a decaying function and the more rapidly it decays, the less memory the system has of distantly occurring excitations. Indeed the characteristic time of this relaxation may be regarded as the "memory time" of the system. Subsequent discussion will relate the step response function to the autocorrelation function of the response variable. Thus the memory time is also the correlation time, introduced in Sec. 5.1.4.

It is often observed that the relaxation of the step response function follows (at least approximately) a decaying exponential

$$\Phi(t) = \chi_0 e^{-t/\tau} \tag{5.60}$$

since we know that the zero-time value of the step response function is the static susceptibility. In this section we shall explore the consequences of this.

The dynamical response function is minus the derivative of $\Phi(t)$

$$X(t) = \chi_0 \frac{1}{t} e^{-t/\tau}.$$

And the frequency domain susceptibility is given by the Fourier transform. Thus

$$\chi(\omega) = \frac{\chi_0}{1 - i\omega\tau}, \tag{5.61}$$

its real and imaginary parts being given by

$$\chi'(\omega) = \chi_0 \frac{1}{1 + \omega^2 \tau^2}$$

$$\chi''(\omega) = \chi_0 \frac{\omega\tau}{1 + \omega^2 \tau^2}. \tag{5.62}$$

It is these functions that are plotted in Fig. 5.14.

This is often called the Debye form for the dynamical susceptibility. The electric susceptibility of many dielectrics follows this form to a *reasonable* extent.

5.5. Linear Response — Microscopics

5.5.1. *Onsager's hypothesis*

We have discussed many formal aspects of the dynamical response function $X(t)$ and its Fourier transform, the dynamical susceptibility $\chi(\omega)$. These provide a description for the evolution of a system from a non-equilibrium state, following a disturbance and they also give the energy dissipation under the application of a such disturbance. The formalism is quite elegant, but it does not tell us how to calculate any of these from microscopic first principles. In this respect, it is more of the nature of thermodynamics rather than of statistical mechanics.

It is possible to calculate such response functions from microscopic first principles, certainly in the linear regime, using statistical–mechanical perturbation theory. However such techniques are, unfortunately, fairly complicated; they require methods beyond the scope of this book. The interested reader is referred particularly to the works of Ryogo Kubo.[9] We shall give a flavour of these methods in Secs. 5.5.3 and 5.5.4 below.

Fortunately there is another way, perhaps a more intuitive way, of tackling the question; this was pioneered by Lars Onsager. Imagine looking at the equilibrium fluctuations in a system, perhaps observing through a microscope the fluctuations in a fluid. The observed quantity will vary in some random way about its mean value. Sometimes there will be small excursions and sometimes, large excursions from the mean. These excursions return, on the average, to the mean value. Now imagine that someone has applied a disturbance to this system, driving the system from equilibrium, and then the disturbance is removed. The system will then return to its equilibrium state. Through our microscope we would see the observed quantity fluctuating while it returns, on average, to its equilibrium value. Onsager made the remarkable hypothesis[10] that one would not be able to distinguish between the two situations. In other words, he assumed that the relaxation of a system following a disturbance is the same as the average regression of a fluctuation in an equilibrium system.

This is indeed remarkable. It says that the behaviour of a *non-equilibrium* system may be understood by studying the properties of the corresponding *equilibrium* system. Alternatively, one may say that at the *microscopic* level there is no distinction between equilibrium and non-equilibrium. But then, of course, equilibrium is a thoroughly *macroscopic* concept.

The point, for us, is that we already know the average way fluctuations, in an equilibrium system, regress towards the mean. This is given by the autocorrelation function of the fluctuating quantity, $\langle M(0)M(t)\rangle$, discussed and explained in Secs. 5.1.2 and 5.1.3. The relaxation following the removal of a disturbance is described by the step response function $\Phi(t)$. Thus Onsager's hypothesis may be expressed as

$$\Phi(t) = \beta\langle M(0)M(t)\rangle \qquad (5.63)$$

where β is a constant to be determined. From the considerations of Sec. 5.4.8 we know that β has the dimensions of inverse energy.

A full microscopic calculation is needed to determine the value of β. For inspiration, we shall look at the zero-time value of Eq. (5.63). Since $\Phi(0)$ is equal to the static susceptibility χ_0, we have

$$\chi_0 = \beta\langle M^2\rangle.$$

Compare this with the magnetic case, expressed in Eq. (5.59). We are thus led to identify β with $1/kT$. It is *plausible* that this result is more generally applicable; we assert that it is so.

The equation for the step response function

$$\Phi(t) = \frac{1}{kT}\langle M(0)M(t)\rangle,$$

or, equivalently for the dynamical response function

$$X(t) = -\frac{1}{kT}\frac{d}{dt}\langle M(0)M(t)\rangle \qquad (5.64)$$

is indeed the correct *classical* result, as could be derived using statistical–mechanical perturbation theory. A quantum-mechanical calculation gives a more complicated temperature dependence, which reduces to the above expression in the high temperature limit. The kT factor is essentially a reflection of equipartition.

The dynamical susceptibility is then found from the Fourier transform of $X(t)$:

$$\chi(\omega) = -\frac{1}{kT}\int_0^\infty \langle M(0)\dot{M}(t)\rangle e^{i\omega t}\,dt \qquad (5.65)$$

or, upon integration by parts

$$\chi(\omega) = \chi_0 + \frac{i\omega}{kT}\int_0^\infty \langle M(0)M(t)\rangle e^{i\omega t}\,dt.$$

This shows how the dynamical susceptibility depends on the autocorrelation function of the equilibrium fluctuations in the response variable.

5.5.2. *Nyquist's theorem*

As early as 1906, Einstein[11] predicted that the random motion of charge carriers in a conductor would lead to fluctuations of voltage or current in any system in thermal equilibrium. This was first observed by Johnson[12] and subsequently explained theoretically by Nyquist,[13] who calculated the power spectrum of the fluctuations.

In order to frame this treatment within the discussion of the previous sections, the generalised force will now be taken as the applied voltage V and the generalised response will be the electrical charge Q. Then the linear response relation between these two will be

$$Q(t) = \int X(t - \tau)V(t)\mathrm{d}\tau$$

or, in the frequency domain

$$q(\omega) = \chi(\omega)v(\omega).$$

Thus in electrical terms we identify the generalised susceptibility in this case as a frequency-dependent capacitance. Then the imaginary part of this will be related to conductance and thus dissipation. Recall we saw in Sec. 5.4.8 that the imaginary part of the generalised susceptibility was connected with dissipation.

The admittance Y is given by $i\omega$ multiplied by the capacitance, so that from Eq. (5.65), this is

$$Y(\omega) = -\frac{i\omega}{kT} \int_0^\infty \langle Q(0)\dot{Q}(t)\rangle e^{i\omega t}\mathrm{d}t.$$

Note, here, that $\dot{Q}(t)$ is the electric current $I(t)$. Let us integrate the above equation by parts in such a way as to differentiate the other Q. This will give

$$Y(\omega) = \frac{1}{kT} \int_0^\infty \langle I(0)I(t)\rangle e^{i\omega t}\mathrm{d}t.$$

This result tells us that the electrical admittance of an object is related to the equilibrium current fluctuations in it. Indeed we can decompose the admittance into its real and imaginary parts: the conductance G and the susceptance S

$$Y(\omega) = G(\omega) + iS(\omega)$$

so that

$$G(\omega) = \frac{1}{kT} \int_0^\infty \langle I(0)I(t) \rangle \cos \omega t \, dt$$

$$S(\omega) = \frac{1}{kT} \int_0^\infty \langle I(0)I(t) \rangle \sin \omega t \, dt.$$

The expression for $G(\omega)$ is a generalisation of Eq. (5.31) derived through the Langevin equation discussion.

Nyquist's result follows from inverting the expression for the conductance $G(\omega)$

$$\langle I(0)I(t) \rangle = kT \frac{2}{\pi} \int_0^\infty G(\omega) \cos \omega t \, d\omega.$$

If we now let t tend to zero, then we obtain the mean square value for the current fluctuations

$$\langle I^2 \rangle = \frac{2kT}{\pi} \int_{-\infty}^\infty G(\omega) d\omega. \tag{5.66}$$

Thus, in a frequency band $\Delta f = \Delta \omega / 2\pi$, there will be a contribution to the mean square current of

$$\langle I^2 \rangle_{\Delta f} = 4kTG\Delta f. \tag{5.67}$$

Alternatively, in a resistance of $R = 1/G$, there will be mean square voltage fluctuations of

$$\langle V^2 \rangle_{\Delta f} = 4kTR\Delta f. \tag{5.68}$$

This so-called "Johnson noise" is commonly heard as the hiss from the loudspeakers of a hi-fi amplifier when there is no signal output.

We first encountered Johnson noise as the thermal equilibrium black body radiation in the (one-dimensional) terminated transmission line, in Sec. 2.6.5. That treatment obtained the full quantum-mechanical result, of which Eqs. (5.67) and (5.68) are the high temperature/equipartition limit. We next saw it from the perspective of the electrical analogue of the Langevin equation, in Sec. 5.3.6.

5.5.3. *Calculation of the step response function*

We introduced the step response function in Sec. 5.4.4. We imagined a constant excitation $B(t) = b$ to have been applied back into the past, and we remove this at time $t = 0$. For negative times the response $M(t)$ will be a constant. And interest focuses on how $M(t)$ relaxes to its new equilibrium value when B drops to zero.

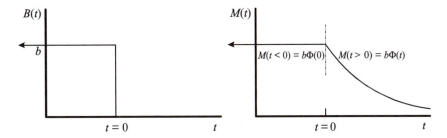

Fig. 5.15.   Step response.

The step response function $\Phi(t)$ is then defined, for $t > 0$, as

$$\Phi(t) = \frac{1}{b} M(t). \tag{5.69}$$

And the question is then to (try and) calculate this from first principles. In this section, we shall show how $\Phi(t)$ is related to the autocorrelation function of $M(t)$, obtaining the Onsager expression. By no means can the derivation be regarded as rigorous, but it is hoped that the flavour of the procedure will be appreciated. The question of how the autocorrelation function may be calculated will be considered in the following section.

The mean value of the response $M(t)$ is found by evaluating the average, weighted by the appropriate Boltzmann factor:

$$\langle M(t) \rangle = \frac{1}{Z} \sum_i M_i(t) e^{-E_i/kT}.$$

Now the system has been prepared by applying the "field" b for negative times. This has the effect of altering the energy levels by adding the term bM_i. Then in calculating the mean values we must use the appropriately-modified Boltzmann factor

$$\langle M(t) \rangle = \frac{1}{Z} \sum_i M_i(t) e^{-(E_i + bM_i)/kT}.$$

So the step response function is then

$$\Phi(t) = \frac{1}{b}\frac{1}{Z}\sum_i M_i(t)e^{-(E_i+bM_i)/kT}.$$

The point about *linear* response is that we take the excitation b to be very small. In this case, we can make the approximation

$$e^{-(E_i+bM_i)/kT} \sim e^{-E_i/kT}(1 + bM_i/kT).$$

And then

$$\Phi(t) = \frac{1}{b}\frac{1}{Z}\sum_i (1 + bM_i/kT)M_i(t)e^{-E_i/kT}$$

$$= \frac{1}{b}\frac{1}{Z}\sum_i M_i(t)e^{-E_i/kT} + \frac{1}{Z}\sum_i (M_i/kT)M_i(t)e^{-E_i/kT},$$

which may be written as

$$\Phi(t) = \frac{1}{b}\langle M(t)\rangle + \frac{1}{kT}\langle M(0)M(t)\rangle.$$

Here the averages are evaluated over the *equilibrium* ensemble. The first term is the equilibrium value of the response — in the absence of any excitation. We may take this to be zero. The second term is the linear response to the excitation. Observe that b has cancelled, indicating that it is indeed a linear response. Here we have taken M from the shifted energy to be the zero-time value, as it has this value for all negative times. Thus we conclude that the step response function, in the linear regime is given by

$$\Phi(t) = \frac{1}{kT}\langle M(0)M(t)\rangle, \qquad (5.70)$$

as proposed in Sec. 5.4.10.

The important point about this result — we have mentioned it before but it is worth repeating it — is that the relaxation from a non-equilibrium state is expressed in terms of the fluctuations in the equilibrium state.

5.5.4. *Calculation of the autocorrelation function*

The calculation of the dynamical susceptibility from microscopic first principles is clearly a very difficult problem. As with our introductory discussion of the foundations of statistical mechanics, some sort of statistical treatment will be needed to solve any real problem. Moreover, almost all calculational procedures will involve the use of approximations.

There are some general methods of approach, often using a Boltzmann equation, to calculate the evolution of a system in phase space from specified initial conditions. Unfortunately this is beyond the scope of this book. However we note that the discussions in Sec. 5.4 impose various constraints on any autocorrelation function so-calculated. In terms of the dynamical susceptibility $X(t)$ and its Fourier transform, the frequency-dependent susceptibility $\chi(\omega)$, we know that:

- causality places restrictions on the relationship between the real part $\chi'(\omega)$ and the imaginary part $\chi''(\omega)$;
- energy considerations restrict $\chi''(\omega)$ to be positive for all (positive) frequencies;
- the reality of $X(t)$ requires that $\chi''(0)$ is zero;
- causality and microscopic reversibility require that the first derivative $\Phi'(0)$ is zero; the reason for this is explored in Problem 5.3.

So any approximately calculated susceptibility must satisfy these requirements.

Hydrodynamic considerations will often give further information; the long-time behaviour of $X(t)$ and the low-frequency behaviour of $\chi(\omega)$ will have further constraints imposed upon them. The way in which such hydrodynamic constraints may be incorporated into the description of correlation functions has been extensively explored by Kadanoff and Martin.[14]

In a very few cases a full quantum-mechanical description of the dynamics might be appropriate. Spin diffusion in a paramagnet is a classical example of this.[15] It is then possible to calculate the short-time behaviour of $X(t)$ as a power series in time. Higher-order derivatives become increasingly difficult to evaluate, but knowledge of the first few derivatives may be combined with long-time hydrodynamic information through the use of an appropriate interpolation procedure. The reader is referred to the author's book on *Nuclear Magnetic Resonance*,[16] particularly the section dealing with solid ^{3}He, for further details. Section 6.3.7 of that book also shows how short-time expansions actually place constraints on the behaviour of the correlation functions at long times.

Problems

5.1. A random quantity has an exponential autocorrelation function $G(t) = G(0)e^{-\gamma t}$. Calculate its correlation time using the usual definition.

5.2. Show that the autocorrelation function of a periodically varying quantity $m(t) = m \cos \omega t$ is given by

$$G(t) = \frac{m^2}{2} \cos \omega t.$$

Show that the autocorrelation function is independent of the *phase* of $m(t)$. In other words, show that if $m(t) = m \cos(\omega t + \varphi)$, then $G(t)$ is independent of φ.

5.3. The dynamical response function $X(t)$ must vanish at zero times, as shown in Fig. 5.13. What is the physical explanation of this? What is the consequence for the step response function $\Phi(t)$? Is this compatible with an exponentially decaying $\Phi(t)$?

5.4. In Sec. 5.3 we examined the form of the dynamical susceptibility $\chi(\omega)$ that followed from the assumption that the step response function $\Phi(t)$ decayed exponentially. In this question consider a step response function that decays with a Gaussian profile, $\Phi(t) = \chi_0 e^{-t^2/2\tau^2}$. Evaluate the real and imaginary parts of the dynamical susceptibility and plot them as a function of frequency. The real part of the susceptibility is difficult to evaluate without a symbolic mathematics system such as *Mathematica*. Compare and discuss the differences and similarities between this susceptibility and that deduced from the exponential step response function (Debye susceptibility).

5.5. The Debye form for the dynamical susceptibility is

$$\chi'(\omega) = \chi_0 \frac{1}{1 + \omega^2 \tau^2}$$

$$\chi''(\omega) = \chi_0 \frac{\omega \tau}{1 + \omega^2 \tau^2}.$$

Plot the real part against the imaginary part and show that the figure corresponds to a semicircle. This is known as a Cole–Cole plot.

5.6. Plot the Cole–Cole plot (Problem 5.5) for the dynamical susceptibility considered in Problem 5.4. How does it differ from that of the Debye susceptibility.

5.7. The full quantum-mechanical calculation of the Johnson noise of a resistor gives

$$\langle v^2 \rangle_{\Delta f} = 4R \frac{hf}{e^{hf/kT} - 1} \Delta f.$$

Show that this reduces to the classical Nyquist expression at low frequencies. At what frequency will there start to be serious deviations from the Nyquist value? Estimate the value of this frequency.

5.8. Show that for the Debye susceptibility, the relation

$$\chi_0 = \frac{1}{\pi} \int_{-\infty}^{\infty} \frac{\chi''(\omega)}{\omega} \, d\omega$$

holds. Demonstrate that χ'' vanishes sufficiently fast at $\omega = 0$ so there is no pole in the integral and there is thus no need to take the principal part of the integral in the Kramers–Kronig relations.

5.9. In Sec. 5.3.6 we considered an electrical analogue of the Langevin Equation based on a circuit comprising an inductor and a resistor. In this problem we shall examine a different analogue: a circuit of a capacitor and a resistor. Show that the equation analogous to the Langevin equation, in this case, is

$$C \frac{dV(t)}{dt} + \frac{1}{R} V(t) = I(t).$$

Hence show that the fluctuation–dissipation result relates the resistance to the current fluctuations through

$$\frac{1}{R} = \frac{1}{2kT} \int_{-\infty}^{\infty} \langle I(0) \, I(t) \rangle dt.$$

References

[1] C. J. Adkins, *Equilibrium Thermodynamics*, 3rd ed. (Cambridge University Press, 1983).

[2] D. Kondepundi and I. Prigogine, *Modern Thermodynamics* (John Wiley, 1998).

[3] D. K. C. MacDonald, *Noise and Fluctuations* (John Wiley, 1962).

[4] A. Einstein, *Investigations on the Theory of Brownian Movement* (Dover, 1956).

[5] G. E. Uhlenbeck and L. S. Ornstein, On the theory of the Brownian motion, *Phys. Rev.* **36** (1930) 823; Reprinted in *Selected Papers on Noise and Stochastic Processes*, Ed. N. Wax (Dover: NY, 1954).

[6] M. C. Wang and G. E. Uhlenbeck, On the theory of the Brownian motion II, *Rev. Mod. Phys.* **17** (1945) 323; Reprinted in *Selected Papers on Noise and Stochastic Processes*, Ed. N. Wax (Dover: NY, 1954).

[7] B. Cowan, *Classical Mechanics* (Routledge and Kegan Paul, London, 1984).

[8] J. L. Doob, The Brownian movement and stochastic equations, *Ann. Math.* **43** (1942) 351; Reprinted in *Selected Papers on Noise and Stochastic Processes*, Ed. N. Wax (Dover: NY, 1954).

[9] R. Kubo, *Some Aspects of the Statistical-Mechanical Theory of Irreversible Processes*, Colorado Summer Institute for Theoretical Physics 1958, p. 181 (Interscience Publishers, 1959); R. Kubo, The fluctuation-dissipation theorem, *Rep. Prog. Phys.* **24** (1966) 255.

[10] L. Onsager, Reciprocal relations in irreversible processes I and II, *Phys. Rev.* **37** (1931) 405; **38** (1931) 2265.

[11] A. Einstein, *Ann. Phys.* **19** (1906) 289; **19** (1906) 371.

[12] J. B. Johnson, *Nature* **119** (1927) 50; *Phys. Rev.* **29** (1927) 367; *Phys. Rev.* **32** (1928) 97.

[13] H. Nyquist, *Phys. Rev.* **29** (1927) 614; *Phys. Rev.* **32** (1928) 110.

[14] L. P. Kadanoff and P. C. Martin, Hydrodynamic equations and correlation functions, *Ann. Phys.* (NY) **24** (1963) 419–469.

[15] B. Cowan and M. Fardis, Direct measurement of spin diffusion from spin relaxation times in solid ^{3}He, *Phys. Rev.* **B44** (1991) 4304–4313.

[16] B. Cowan, *Nuclear Magnetic Resonance and Relaxation* (Cambridge University Press, 1997).

APPENDIXES

Appendix 1
The Gibbs–Duhem Relation

A.1.1. *Homogeneity of the fundamental relation*

The Gibbs–Duhem relation follows from the fact that entropy is an extensive quantity and that it is a function of the other extensive variables of the system. An important mathematical result can be derived *just* from this. We saw that (for a pV system) entropy is a function of the energy, the volume and the number of particles in the system: $S = S(E, V, N)$ and this can, formally, be rearranged to write the energy as a function of S, V and N

$$E = E(S, V, N).$$

This is referred to as the "fundamental relation" for the system.

Now since E, S, V and N are all extensive variables, the fundamental relation function $E = E(S, V, N)$ is homogeneous. In other words, if the size of the system is multiplied by a constant λ then E, S, V and N are all multiplied by the same λ. Thus

$$E(\lambda S, \lambda V, \lambda N) = \lambda E(S, V, N).$$

A.1.2. *The Euler relation*

In order to investigate the consequences of this homogeneity let us differentiate this expression with respect to λ; we find

$$\frac{\partial E(\lambda S, \lambda V, \lambda N)}{\partial(\lambda S)} \frac{\partial(\lambda S)}{\partial \lambda} + \frac{\partial E(\lambda S, \lambda V, \lambda N)}{\partial(\lambda V)} \frac{\partial(\lambda V)}{\partial \lambda} + \frac{\partial E(\lambda S, \lambda V, \lambda N)}{\partial(\lambda N)} \frac{\partial(\lambda N)}{\partial \lambda}$$
$$= E(S, V, N)$$

or

$$\frac{\partial E(\lambda S, \lambda V, \lambda N)}{\partial(\lambda S)}S + \frac{\partial E(\lambda S, \lambda V, \lambda N)}{\partial(\lambda V)}V + \frac{\partial E(\lambda S, \lambda V, \lambda N)}{\partial(\lambda N)}N = E(S, V, N).$$

If we now set the value of λ to unity then the derivatives are simply T, $-p$ and μ so that

$$E = TS - pV + \mu N.$$

This is the Gibbs–Duhem relation[1]; at least it is so called by some authors. A better designation is the "Euler relation", following Callen.[2] This is appropriate because of Euler's interest in the mathematical properties of homogeneous functions.

A.1.3. *A caveat*

We remark, parenthetically, that the student should not be misled by this equation. In learning about thermodynamics one encounters the concept of a function of state, and the fact that heat and work are not functions of state; they depend on path. The Euler relation tells us that there are three contributions to the internal energy: TS, $-pV$ and μN. But we *cannot* identify TS as the heat content, $-pV$ as the work content and μN as the chemical content. In other words, the path-dependent $\int T\mathrm{d}s$ is different from the function of state TS, etc.

A.1.4. *The Gibbs–Duhem relation*

Probably more authentically deserving of the title Gibbs–Duhem relation is the result of combining the Euler relation with the differential form for $\mathrm{d}E$ (the First Law expression), namely

$$\mathrm{d}E = T\mathrm{d}S - p\mathrm{d}V + \mu\mathrm{d}N$$

to give

$$S\mathrm{d}T - V\mathrm{d}p + N\mathrm{d}\mu = 0.$$

This tells us that the intensive variables T, p and μ are not independent, which may be surprising. We understand temperature and pressure as (independent) variables over which we have fairly direct control. And by varying these we can effect heat and volume "flow". Similarly we learned that particles can be induced to flow by varying the chemical potential. Certainly, our "feel" for chemical potential is not so intuitive as that for temperature and pressure, but we are helped by the mathematical analogies with T and p. However, now we see that if we regard T and p as independent variables then μ depends on T and p; it is not independent.

Appendix 2

Thermodynamic Potentials

The discussions in this Appendix draw unashamedly on the treatment of Callen.[2] That discussion is so clear and intuitive; it cannot be improved upon. The reader is most certainly advised to read Callen's book at some stage of his/her studies.

A.2.1. *Equilibrium states*

We have seen that the equilibrium state of an isolated system — characterised by E, V, N — is determined by maximising the entropy S. On the other hand, we know that a purely mechanical system settles down to the equilibrium state that minimises the energy: the state where the forces $F_i = -\partial E/\partial X_i$ vanish. In this section we shall see how to relate these two ideas. And then in the following sections we shall see how to extend the ideas.

By a purely mechanical system we mean one with *no* thermal degrees of freedom. This means no changing of the *populations* of the different quantum states — i.e. at constant entropy. But this should also apply for a thermodynamic system at constant entropy. So we should be able to find the equilibrium state in two ways:

- Maximise the entropy at constant energy,
- Minimise the energy at constant entropy.

That these approaches are equivalent may be seen by considering the form of the $E - S - X$ surface for a system. Here X is some extensive quantity that will vary when the system approaches equilibrium like the energy or the number of particles in one half of the system.

At constant energy the plane $E = E_0$ intersects the $E - S - X$ surface along a line of possible states of the system. We have seen that the equilibrium state — the state of maximum probability — will be the point of maximum entropy: the point P on the curve.

But now consider the same system at constant entropy. The plane $S = S_0$ intersects the $E - S - X$ surface along a different line of possible

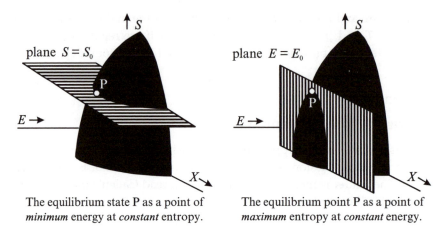

The equilibrium state P as a point of	The equilibrium point P as a point of
minimum energy at *constant* entropy.	*maximum* entropy at *constant* energy.

Fig. A.2.1.   Alternative specification of equilibrium states (after Callen[2]).

states. Comparing the two pictures we see that the equilibrium point P is now the point of minimum energy.

Equivalence of the entropy maximum and the energy minimum principles depends on the shape of the $E - S - X$ surface. In particular, it relies on

(a) S having a maximum with respect to X — guaranteed by the Second Law;
(b) S being an increasing function of E — $\partial S/\partial E = 1/T > 0$ means positive temperatures;
(c) E being a single-valued continuous function of S.

To demonstrate the equivalence of the entropy maximum and the energy minimum principles we shall show that the converse would lead to a violation of the second law.

Assume that the energy E is *not* the minimum value consistent with a given entropy S. Then let us take some energy out of the system in the form of work and return it in the form of heat. The energy is then the same but the entropy will have increased. So the original state could not have been an equilibrium state.

The equivalence of the energy minimum and the entropy maximum principle is rather like describing the circle as having the maximum area at fixed circumference or the minimum circumference for a given area.

We shall now look at the specification of equilibrium states when instead of energy or entropy, other variables are held constant.

A.2.2. *Constant temperature (and volume): the Helmholtz potential*

To maintain a system of fixed volume at constant temperature we shall put it in contact with a heat reservoir. The equilibrium state can be determined by maximising the total entropy while keeping the total energy constant.

The total entropy is the sum of the system entropy and that of the reservoir. The entropy maximum condition is then

$$\left. \begin{array}{r} dS_T = dS + dS_{res} = 0 \\ d^2 S_T < 0. \end{array} \right\}$$

The entropy differential for the reservoir is

$$dS_{res} = \frac{dE_{res}}{T_{res}}$$

$$= -\frac{dE}{T}$$

since the total energy is constant. The total entropy maximum condition is then

$$\left. \begin{array}{r} dS - \dfrac{dE}{T} = 0 \\[2mm] d^2 S - \dfrac{d^2 E}{T} < 0. \end{array} \right\}$$

Or, since T is constant,

$$\left. \begin{array}{r} d(E - TS) = 0 \\ d^2(E - TS) > 0, \end{array} \right\}$$

which is the condition for a minimum in $E - TS$. But we have encountered $E - TS$ before, in the consideration of the link between the partition function and thermodynamic properties. This function is the Helmholtz free energy F. So we conclude that

at constant N, V and T, $F = E - TS$ is a minimum

— The Helmholtz minimum principle.

We can understand this as a competition between two opposing effects. At high temperatures the entropy tends to a maximum, while at low temperatures the energy tends to a minimum. And the balance between these competing processes is given, at general temperatures, by minimising the combination $F = E - TS$.

A.2.3. *Constant pressure and energy: the Enthalpy function*

To maintain a system at constant pressure we shall put it in mechanical contact with a "volume reservoir". That is, it will be connected by a movable, thermally isolated piston, to a very large volume. As before, we can determine the equilibrium state by maximising the total entropy while keeping the total energy constant. Alternatively and equivalently, we can keep the total entropy constant while minimising the total energy.

The energy minimum condition is

$$\left.\begin{array}{r} dE_T = dE + dE_{res} = 0 \\ d^2E_T > 0. \end{array}\right\}$$

In this case, the reservoir may do mechanical work on our system:

$$dE_{res} = -p_{res}\,dV_{res} = pdV$$

since the total volume is fixed. We then write the energy minimum condition as

$$\left.\begin{array}{r} dE + pdV = 0 \\ d^2E + pd^2V > 0, \end{array}\right\}$$

or, since p is constant

$$\left.\begin{array}{r} d(E + pV) = 0 \\ d^2(E + pV) > 0. \end{array}\right\}$$

This is the condition for a minimum in $E + pV$. This function is called the Enthalpy, and it is given the symbol H. So we conclude that

at constant N, p and E, $H = E + pV$ is a minimum
— The Enthalpy minimum principle.

A.2.4. *Constant pressure and temperature: the Gibbs free energy*

In this case, our system can exchange both thermal and mechanical energy with a reservoir; both heat energy and "volume" may be exchanged. Working in terms of the minimum energy at constant entropy condition for the combined system + reservoir

$$dE_T = dE + dE_{res} = 0 \atop d^2 E_T > 0. \Bigg\}$$

In this case, the reservoir gives heat energy and/or it may do mechanical work on our system:

$$dE_{res} = T_{res} dS_{res} - p_{res} dV_{res} = -TdS + pdV$$

since the total energy is fixed. We then write the energy minimum condition as

$$dE - TdS + pdV = 0 \atop d^2 E - Td^2 S + pd^2 V > 0, \Bigg\}$$

or, since T and p are constant

$$d(E - TS + pV) = 0 \atop d^2(E - TS + pV) > 0. \Bigg\}$$

This is the condition for a minimum in $E - TS + pV$. This function is called the Gibbs free energy, and it is given the symbol G. So we conclude that

at constant N, p and T, $G = E - TS + pV$ is a minimum

— The Gibbs free energy minimum principle.

A.2.5. *Differential expressions for the potentials*

The internal energy, Helmholtz free energy, enthalpy and Gibbs free energy are called *thermodynamic potentials*. Clearly they are all functions of state. From the definitions

$$F = E - TS \qquad \text{Helmholtz function}$$
$$H = E + pV \qquad \text{Enthalpy function}$$
$$G = E - TS + pV \quad \text{Gibbs function}$$

and the differential expression for the internal energy

$$dE = TdS - pdV$$

we obtain the differential expressions for the potentials:

$$dF = -SdT - pdV$$
$$dH = TdS + Vdp$$
$$dG = -SdT + Vdp.$$

A.2.6. *Natural variables and the Maxwell relations*

Each of the thermodynamic potentials has its own *natural variables*. For instance, taking E, the differential expression for the first law is

$$dE = TdS - pdV.$$

Thus if E is known as a function of S and V then everything else — like T and p — can be obtained by differentiation, since $T = \partial E/\partial S|_V$ and $p = -\partial E/\partial V|_p$. If, instead, E were known as a function of, say, T and V, then we would need more information, such as an equation of state, to completely determine the remaining thermodynamic functions.

All the potentials have their natural variables in terms of which the dependent variables may be found by differentiation:

$$E(S, V), \quad dE = TdS - pdV, \quad T = \left.\frac{\partial E}{\partial S}\right|_V, \quad p = -\left.\frac{\partial E}{\partial V}\right|_S,$$

$$F(T, V), \quad dF = -SdT - pdV, \quad S = -\left.\frac{\partial F}{\partial T}\right|_V, \quad p = -\left.\frac{\partial F}{\partial V}\right|_T,$$

$$H(S, p), \quad dH = TdS + Vdp, \quad T = \left.\frac{\partial H}{\partial S}\right|_p, \quad V = \left.\frac{\partial H}{\partial p}\right|_S,$$

$$G(T, p), \quad dG = -SdT + Vdp, \quad S = \left.\frac{\partial G}{\partial T}\right|_p, \quad V = -\left.\frac{\partial G}{\partial p}\right|_T.$$

If we differentiate one of these results with respect to a further variable then the order of differentiation is immaterial; differentiation is commutative. Thus, for instance, using the energy natural variables we see that

$$\left.\frac{\partial}{\partial V}\right|_S \left.\frac{\partial}{\partial S}\right|_V = \left.\frac{\partial}{\partial S}\right|_V \left.\frac{\partial}{\partial V}\right|_S,$$

and operating on E with this we obtain

$$\left.\frac{\partial}{\partial V}\right|_S \underbrace{\left.\frac{\partial E}{\partial S}\right|_V}_{\parallel} = \left.\frac{\partial}{\partial S}\right|_V \underbrace{\left.\frac{\partial E}{\partial V}\right|_S}_{\parallel}$$

$$\qquad\qquad T \qquad\qquad\qquad -p$$

so that we obtain the result

$$\left.\frac{\partial T}{\partial V}\right|_S = -\left.\frac{\partial p}{\partial S}\right|_V.$$

Similarly, we get one relation for each potential by differentiating it with respect to its two natural variables

$$E: \quad \left.\frac{\partial T}{\partial V}\right|_S = -\left.\frac{\partial p}{\partial S}\right|_V$$

$$F: \quad \left.\frac{\partial S}{\partial V}\right|_T = \left.\frac{\partial p}{\partial T}\right|_V$$

$$H: \quad \left.\frac{\partial T}{\partial p}\right|_S = \left.\frac{\partial V}{\partial S}\right|_p$$

$$G: \quad \left.\frac{\partial S}{\partial p}\right|_T = -\left.\frac{\partial V}{\partial T}\right|_p.$$

The Maxwell relations give equations between seemingly different quantities. In particular, they often connect easily measured, but uninteresting quantities to difficult-to-measure, but very interesting ones.

Another application of the Maxwell relations is in the study of glassy systems and other quasi-equilibrium states of matter. The validity of the Maxwell relations relies on the assumption of a state of thermal equilibrium. An experimental demonstration that a Maxwell relation does not hold is then an indication that the system under investigation is not in a state of thermal equilibrium.

Appendix 3

Mathematica Notebooks

This Appendix contains various *Mathematica* Notebooks for the calculation of properties of gases in the low- and high-temperature limits. *Mathematica* input lines are indicated by **typewriter bold** text while *Mathematica* output lines are indicated by typewriter normal text.

A.3.1. *Chemical potential of Fermi gas at low temperatures*

This *Mathematica* Notebook evaluates the low temperature chemical potential of a Fermi gas as a power series in temperature using the Sommerfeld method discussed in Sec. 2.4.3.

This approach is based on using the following approximation for the internal energy iterms of the chemical potential

$$E_F^{3/2} = \int_{-\infty}^{\infty} \frac{e^x(kTx+\mu)^{3/2}}{(e^x+1)^2}\, dx.$$

The bracket in the numerator is expanded as a series in powers of temperature and then the integral is evaluated term by term.

Let us first specify the order of the power series expansion n:

n = 4

4

Now define the numerator bracket zz

zz = (kTx + μ) ^ (3/2)

$(kTx + \mu)^{3/2}$

and then evalute its power series expansion

sz = Series[zz, {x,0,n}]

$$\mu^{3/2} + \frac{3}{2}kT\sqrt{\mu}x + \frac{3kT^2x^2}{8\sqrt{\mu}} - \frac{kT^3x^3}{16\mu^{3/2}} \frac{3kT^4x^4}{128\mu^{5/2}} + O[x]^5$$

This is a series object. It must be converted to a regular polynomial before it can be integrated — make it Normal:

nsz = Normal[sz]

$$\frac{3kT^4x^4}{128\mu^{5/2}} + \frac{kT^3x^3}{16\mu^{3/2}} + \frac{3kT^2x^2}{8\sqrt{\mu}} + \frac{3}{2}kTx\sqrt{\mu} + \mu^{3/2}$$

and now we can do the integration

ef32 = Integrate[Exp[x]nsz/(Exp[x]+1)^2,

{x,-Infinity,Infinity}] // Apart

$$\frac{7kT^4\pi^4}{640\mu^{5/2}} + \frac{kT^2\pi^2}{8\sqrt{\mu}} + \mu^{3/2}$$

The expression for the Fermi energy is found by taking the 2/3 power

ef = ef32^(2/3)

$$\left(\frac{7kT^4\pi^4}{640\mu^{5/2}} + \frac{kT^2\pi^2}{8\sqrt{\mu}} + \mu^{3/2} \right)^{2/3}$$

and this can then be expressed as a power series by:

sef = Series[ef, {μ, Infinity, n}]

$$\frac{1}{\frac{1}{\mu}} + \frac{kT^2\pi^2}{12\mu} + \frac{1}{180}kT^4\pi^4\left(\frac{1}{\mu}\right)^3 + 0\left[\frac{1}{\mu}\right]^5$$

This is the series for Ef in terms of μ, which must be inverted to give a series for μ in terms of Ef. However there seems to be a bug in *Mathematica* so that it is necessary to make the series Normal and then turn it into another series before inverting it.

nef = Normal[sef]

$$\frac{kT^4\pi^4}{180\mu^3} + \frac{kT^2\pi^2}{12\mu} + \mu$$

snef = Series[nef, {μ, Infinity, n}]

$$\frac{1}{\frac{1}{\mu}} + \frac{kT^2\pi^2}{12\mu} + \frac{1}{180}kT^4\pi^4\left(\frac{1}{\mu}\right)^3 + 0\left[\frac{1}{\mu}\right]^5$$

The next command inverts the series, giving μ in terms of the Fermi energy.

smu = InverseSeries[snef, Ef]

$$\frac{1}{\frac{1}{Ef}} - \frac{kT^2\pi^2}{12Ef} - \frac{1}{80}(kT^4\pi^4)\left(\frac{1}{Ef}\right)^3 + 0\left[\frac{1}{Ef}\right]^5$$

mu = Normal[smu]

$$Ef - \frac{kT^2\pi^2}{12Ef} - \frac{kT^4\pi^4}{80Ef^3}$$

This is the required low temperature series expression for the chemical potential.

A.3.2. *Internal energy of the Fermi gas at low temperatures*

This *Mathematica* Notebook evaluates the low temperature internal energy of a Fermi gas as a power series in temperature using the Sommerfeld method discussed in Sec. 2.4.3. It uses the series for the chemical potential obtained in the previous Notebook.

This approach is based on using the following approximation for the internal energy in terms of the chemical potential

$$E = \frac{3}{5}\frac{N}{E_F^{3/2}}\int_{-\infty}^{\infty}\frac{e^x(kTx+\mu)^{5/2}}{(e^x+1)^2}\,dx$$

The bracket in the denominator is expanded as a series in powers of temperature and then the integral is evaluated term by term.

Let us first specify the order of the power series expansion n:

n = 4

4

Now define the numerator bracket zz

zz = (kTx + μ)^(5/2)

$(kTx + \mu)^{5/2}$

and then evalute its power series expansion

sz = Series[zz, {x,0,n}]

$$\mu^{5/2} + \frac{5}{2}kT\mu^{3/2}x + \frac{15}{8}kT^2\sqrt{\mu}x^2 + \frac{5kT^3x^3}{16\sqrt{\mu}} - \frac{5kT^4x^4}{128\mu^{3/2}} + O[x]^5$$

This is a series object. It must be converted to a regular polynomial before it can be integrated — make it Normal:

nsz = Normal[sz]

$$-\frac{5kT^4x^4}{128\mu^{3/2}} + \frac{5kT^3x^3}{16\sqrt{\mu}} + \frac{15}{8}kT^2x^2\sqrt{\mu} + \frac{5}{2}kTx\mu^{3/2} + \mu^{5/2}$$

and now we can do the integration

enoNoef32 = (3/5) Integrate[Exp[x]nsz/(Exp[x]+1)^2,
{x,-Infinity,Infinity}] // Apart

$$-\frac{7kT^4\pi^4}{640\mu^{3/2}} + \frac{3}{8}kT^2\pi^2\sqrt{\mu} + \frac{3\mu^{5/2}}{5}$$

Now we substitute in for the chemical potential, from a previous calculation

$$\mu = \mathbf{Ef} - \frac{kT^2\pi^2}{12Ef} - \frac{kT^4\pi^4}{80Ef^3}$$

$$Ef - \frac{kT^2\pi^2}{12Ef} - \frac{kT^4\pi^4}{80Ef^3}$$

enoNoef32

$$-\frac{7\mathrm{kT}^4\pi^4}{640\left(\mathrm{Ef}-\frac{\mathrm{kT}^2\pi^2}{12\mathrm{Ef}}-\frac{\mathrm{kT}^4\pi^4}{80\mathrm{Ef}^3}\right)^{3/2}}+\frac{3}{8}\mathrm{kT}^2\pi^2\sqrt{\mathrm{Ef}-\frac{\mathrm{kT}^2\pi^2}{12\mathrm{Ef}}-\frac{\mathrm{kT}^4\pi^4}{80\mathrm{Ef}^3}}$$

$$+\frac{3}{5}\left(\mathrm{Ef}-\frac{\mathrm{kT}^2\pi^2}{12\mathrm{Ef}}-\frac{\mathrm{kT}^4\pi^4}{80\mathrm{Ef}^3}\right)^{5/2}$$

energyoN = enoNoef32/Ef^(3/2)

$$\frac{1}{\mathrm{Ef}^{3/2}}\left(-\frac{7\mathrm{kT}^4\pi^4}{640\left(\mathrm{Ef}-\frac{\mathrm{kT}^2\pi^2}{12\mathrm{Ef}}-\frac{\mathrm{kT}^4\pi^4}{80\mathrm{Ef}^3}\right)^{3/2}}\right.$$

$$\left.+\frac{3}{8}\mathrm{kT}^2\pi^2\sqrt{\mathrm{Ef}-\frac{\mathrm{kT}^2\pi^2}{12\mathrm{Ef}}-\frac{\mathrm{kT}^4\pi^4}{80\mathrm{Ef}^3}}+\frac{3}{5}\left(\mathrm{Ef}-\frac{\mathrm{kT}^2\pi^2}{12\mathrm{Ef}}-\frac{\mathrm{kT}^4\pi^4}{80\mathrm{Ef}^3}\right)^{5/2}\right)$$

Series[energyoN, {kT,0,n}]

$$\frac{3\mathrm{Ef}}{5}+\frac{\pi^2\mathrm{kT}^2}{4\mathrm{Ef}}-\frac{3\pi^4\mathrm{kT}^4}{80\mathrm{Ef}^3}+\mathrm{O[kT]}^5$$

This is the required low temperature series expression for the internal energy.

A.3.3. *Fugacity of the ideal gas at high temperatures — Fermi, Maxwell and Bose cases*

This *Mathematica* Notebook evaluates the high temperature fugacity (and thus the chemical potential) of the Fermi, Maxwell and Bose gases.

This approach is based on using the following approximation for the Fermi energy in terms of the chemical potential

$$\varepsilon_F^{3/2}=\frac{3}{2}\int_0^\infty\frac{\varepsilon^{1/2}}{e^{(\varepsilon-\mu)/kT}+a}\mathrm{d}\varepsilon$$

where a is $+1$ for fermions, 0 for maxwellons and -1 for bosons.

We write $x = e^{(\varepsilon-\mu)/kT}$ in order to expand the numerator in powers of this.

Let us first specify the order of the power series expansion n:

n = 4

4

Now define the denominator factor de

de = 1/ (1 + ax)

$$\frac{1}{1 + ax}$$

and we expand this as a series

sde = Series [de, {x, 0, n}]

$1 - ax + a^2x^2 - a^3x^3 + a^4x^4 + O[x]^5$

ar is the argument of the integral for the Fermi energy. We introduce the fugacity z also:

ar = (e^(1/2) x Normal[sde]) /. x → z Exp[(-e)/kT]

$$\sqrt{e}\, e^{-\frac{e}{kT}}z\left(1 - ae^{-\frac{e}{kT}}z + a^2e^{-\frac{2e}{kT}}z^2 - a^3e^{-\frac{3e}{kT}}z^3 + a^4e^{-\frac{4e}{kT}}z^4\right)$$

And now we integrate to find the 3/2 power of the Fermi energy

Ef32 = (3/2) Integrate [ar, {e, 0, Infinity},

　　　Assumptions → Re[kT] > 0]

$$\frac{kT^{3/2}\sqrt{\pi}z(1800 + az(-450\sqrt{2} + az(200\sqrt{3} + 9az(-25 + 8\sqrt{5}az))))}{2400}$$

Apart [Ef32]

$$\frac{3}{4}kT^{3/2}\sqrt{\pi}z - \frac{3}{8}akT^{3/2}\sqrt{\frac{\pi}{2}}z^2 + \frac{1}{4}a^2kT^{3/2}\sqrt{\frac{\pi}{3}}z^3 - \frac{3}{32}a^3kT^{3/2}\sqrt{\pi}z^4$$

$$+ \frac{3}{20}a^4kT^{3/2}\sqrt{\frac{\pi}{5}}z^5$$

Ef = Ef32^(2/3)

$$\frac{\left(\frac{\pi}{10}\right)^{1/3}(kT^{3/2}z(1800 + az(-450\sqrt{2} + az(200\sqrt{3} + 9az(-25 + 8\sqrt{5}az)))))^{2/3}}{40^{2/3}}$$

Here is the series expression for the Fermi energy

`SEf = Series[Ef, {z,0,n}]`

$$\frac{1}{2}3^{2/3}(kT^{3/2})^{2/3}\left(\frac{\pi}{2}\right)^{1/3}z^{2/3} - \frac{\left(a(kT^{3/2})^{2/3}\left(\frac{\pi}{3}\right)^{1/3}\right)z^{5/3}}{22^{5/6}}$$

$$+\left(-\frac{1}{48}a^2(kT^{3/2})^{2/3}\left(\frac{\pi}{6}\right)^{1/3}+\frac{a^2(kT^{3/2})^{2/3}\left(\frac{\pi}{2}\right)^{1/3}}{33^{5/6}}\right)z^{8/3}$$

$$+\left(-\frac{1}{8}a^3(kT^{3/2})^{2/3}\left(\frac{\pi}{6}\right)^{1/3}-\frac{a^3(kT^{3/2})^{2/3}\left(\frac{\pi}{3}\right)^{1/3}}{2162^{5/6}}\right.$$

$$\left.+\frac{a^3(kT^{3/2})^{2/3}\pi^{1/3}}{186^{5/6}}\right)z^{11/3}+O[z]^{14/3}$$

We now invert the series to find the fugacity z in powers of the Fermi energy

`zz = InverseSeries[SEf, eF]`

$$\frac{4\left(\frac{(kT^{3/2})^{1/3}}{kT^{3/2}}\right)^{3/2}eF^{3/2}}{3\sqrt{\pi}}$$

$$+\frac{4\sqrt{2}aeF^3}{9kT^3\pi}+\frac{4\left(\frac{(kT^{3/2})^{1/3}}{kT^{3/2}}\right)^{3/2}\left(\frac{2a^2}{27kT^3\pi}+\frac{2(15a^2-8\sqrt{3}a^2)}{81kT^3\pi}\right)eF^{9/2}}{3\sqrt{\pi}}$$

$$+\frac{1}{3\sqrt{\pi}}\left(4\left(\frac{(kT^{3/2})^{1/3}}{kT^{3/2}}\right)^{3/2}\left(\frac{a(kT^{3/2})^{2/3}\left(\frac{2a^2}{81kT^3\pi}+\frac{4(15a^2-8\sqrt{3}a^2)}{243kT^3\pi}\right)\sqrt{\frac{2}{\pi}}}{9kT^3\sqrt{\frac{(kT^{3/2})^{1/3}}{kT^{3/2}}}}\right.\right.$$

$$+\frac{8\sqrt{2}a(15a^2-8\sqrt{3}a^2)(kT^{3/2})^{2/3}}{2187kT^6\sqrt{\frac{(kT^{3/2})^{1/3}}{kT^{3/2}}}\pi^{3/2}}$$

$$\left.\left.+\frac{8(27a^3+26\sqrt{2}a^3-26\sqrt{6}a^3)(kT^{3/2})^{2/3}}{729kT^6\sqrt{\frac{(kT^{3/2})^{1/3}}{kT^{3/2}}}\pi^{3/2}}\right)eF^6\right)+O[eF]^{15/2}$$

In order to simplify this series expression we must first make it "normal"

```
nz = Normal[zz]
```

$$\frac{4\sqrt{2}aeF^3}{9kT^3\pi} + \frac{4eF^{3/2}\left(\frac{(kT^{3/2})^{1/3}}{kT^{3/2}}\right)^{3/2}}{3\sqrt{\pi}} + \frac{1}{3\sqrt{\pi}}$$

$$\times \left(4eF^6\left(\frac{(kT^{3/2})^{1/3}}{kT^{3/2}}\right)^{3/2}\left(\frac{a(kT^{3/2})^{2/3}\left(\frac{2a^2}{81kT^3\pi} + \frac{4(15a^2 - 8\sqrt{3}a^2)}{243kT^3\pi}\right)\sqrt{\frac{2}{\pi}}}{9kT^3\sqrt{\frac{(kT^{3/2})^{1/3}}{kT^{3/2}}}}\right.\right.$$

$$+ \frac{8\sqrt{2}a(15a^2 - 8\sqrt{3}a^2)(kT^{3/2})^{2/3}}{2187kT^6\sqrt{\frac{(kT^{3/2})^{1/3}}{kT^{3/2}}}\pi^{3/2}}$$

$$\left.\left.+ \frac{8(27a^3 + 26\sqrt{2}a^3 - 26\sqrt{6}a^3)(kT^{3/2})^{2/3}}{729kT^6\sqrt{\frac{(kT^{3/2})^{1/3}}{kT^{3/2}}}\pi^{3/2}}\right)\right)$$

$$+ \frac{4eF^{9/2}\left(\frac{(kT^{3/2})^{1/3}}{kT^{3/2}}\right)^{3/2}\left(\frac{2a^2}{27kT^3\pi} + \frac{2(15a^2 - 8\sqrt{3}a^2)}{81kT^3\pi}\right)}{3\sqrt{\pi}}$$

and then expand as a series in inverse powers of temperature

```
snz = Series[nz, {kT, Infinity, n}]
```

$$\frac{4eF^{3/2}\left(\frac{1}{kT}\right)^{3/2}}{3\sqrt{\pi}} + \frac{4\sqrt{2}aeF^3\left(\frac{1}{kT}\right)^3}{9\pi} + O\left[\frac{1}{kT}\right]^{9/2}$$

This is the result we seek, the fugacity z in powers of 1/T
    We find the classical fugacity by setting $a = 0$.

```
classicalz = Normal[snz /. a → 0]
```

$$\frac{4eF^{3/2}\left(\frac{1}{kT}\right)^{3/2}}{3\sqrt{\pi}}$$

The Fermi fugacity is found by setting $a = 1$

```
fermiz = Normal[snz /. a → 1]
```

$$\frac{4\sqrt{2}eF^3}{9kT^3\pi} + \frac{4eF^{3/2}\left(\frac{1}{kT}\right)^{3/2}}{3\sqrt{\pi}}$$

And the Bose fugacity is found by setting $a = -1$

```
bosez = Normal [snz /. a → -1]
```

$$-\frac{4\sqrt{2}eF^3}{9kT^3\pi} + \frac{4eF^{3/2}\left(\frac{1}{kT}\right)^{3/2}}{3\sqrt{\pi}}$$

A.3.4. *Internal energy of the ideal gas at high temperatures — Fermi, Maxwell and Bose cases*

This *Mathematica* Notebook evaluates the high temperature internal energy of the Fermi, Maxwell and Bose gases.

This approach is based on using the following approximation for the internal energy in terms of the chemical potential

$$E = \frac{3N}{2\varepsilon_F} \int_0^\infty \frac{\varepsilon^{3/2}}{e^{(\varepsilon-\mu)/kT} + a} \, d\varepsilon$$

where a is $+1$ for fermions, 0 for maxwellons and -1 for bosons. The chemical potential (fugacity) is imported from a previous *Mathematica* Notebook as a similar high temperature series.

We write $x = e^{(\varepsilon-\mu)/kT}$ in order to expand the numerator in powers of this.

Let us first specify the order of the power series expansion n:

```
n = 4
```

4

Now define the denominator factor de

```
de = 1/ (1 + ax)
```

$$\frac{1}{1 + ax}$$

and we expand this as a series

```
sde = Series [de, {x, 0, n}]
```

$$1 - ax + a^2x^2 - a^3x^3 + a^4x^4 + O[x]^5$$

ar is the argument of the integral for the internal energy. We introduce the fugacity z also:

```
ar = (e^(3/2) x Normal [sde] N/eF^(3/2)) /. x → z
Exp [ (-e) /kT]
```

$$\frac{e^{3/2}e^{-\frac{e}{kT}}Nz\left(1 - ae^{-\frac{e}{kT}}z + a^2e^{-\frac{2e}{kT}}z^2 - a^3e^{-\frac{3e}{kT}}z^3 + a^4e^{-\frac{4e}{kT}}z^4\right)}{eF^{3/2}}$$

And now we integrate to find the internal energy EE

```
EE = (3/2) Integrate[ar, {e, 0, Infinity},
     Assumptions → Re4[kT] > 0]
```

$$\frac{kT^{5/2}N\sqrt{\pi}z(108000+az(-13500\sqrt{2}+az(4000\sqrt{3}+27az(-125+32\sqrt{5}az))))}{96000eF^{3/2}}$$

```
Apart[EE]
```

$$\frac{9kT^{5/2}N\sqrt{\pi}z}{8eF^{3/2}} - \frac{9akT^{5/2}N\sqrt{\frac{\pi}{2}}z^2}{32eF^{3/2}} + \frac{a^2kT^{5/2}N\sqrt{\frac{\pi}{3}}z^3}{8eF^{3/2}}$$

$$-\frac{9a^3kT^{5/2}N\sqrt{\pi}z^4}{256eF^{3/2}} + \frac{9a^4kT^{5/2}N\sqrt{\frac{\pi}{5}}z^5}{200eF^{3/2}}$$

We now import the series expression for the fugacity

$$zz = \frac{4eF^{3/2}\left(\frac{1}{kT}\right)^{3/2}}{3\sqrt{\pi}} + \frac{4\sqrt{2}aeF^3\left(\frac{1}{kT}\right)^3}{9\pi}$$

$$\frac{4\sqrt{2}aeF^3}{9kT^3\pi} + \frac{4eF^{3/2}\left(\frac{1}{kT}\right)^{3/2}}{3\sqrt{\pi}}$$

```
EEE = Apart[Simplify[EE /. z → zz,
      Assumptions → Re[kT]>0]]
```

$$\frac{3kTN}{2} + \frac{512\sqrt{\frac{2}{5}}a^9eF^{27/2}N}{164025kT^{25/2}\pi^{9/2}} + \frac{512a^8eF^{12}N}{10935\sqrt{5}kT^{11}\pi^4}$$

$$+ \frac{4(-25a^7eF^{21/2}N + 128\sqrt{10}a^7eF^{21/2}N)}{18225kT^{19/2}\pi^{7/2}}$$

$$- \frac{8(25\sqrt{2}a^6eF^9N - 64\sqrt{5}a^6eF^9N)}{6075kT^8\pi^3}$$

$$+ \frac{4(-2025a^5eF^{15/2}N + 100\sqrt{6}a^5eF^{15/2}N + 864\sqrt{10}a^5eF^{15/2}N)}{54675kT^{13/2}\pi^{5/2}}$$

$$- \frac{4(1125\sqrt{2}a^4eF^6N - 500\sqrt{3}a^4eF^6N - 288\sqrt{5}a^4eF^6N)}{30375kT^5\pi^2}$$

$$+ \frac{-18a^3eF^{9/2}N - 9\sqrt{2}a^3eF^{9/2}N + 16\sqrt{6}a^3eF^{9/2}N}{162kT^{7/2}\pi^{3/2}}$$

$$+ \frac{-27a^2eF^3N + 8\sqrt{3}a^2eF^3N}{81kT^2\pi} + \frac{aeF^{3/2}N}{2\sqrt{kT}\sqrt{2\pi}}$$

Here is the series expression for the internal energy

```
SEEE = Series[EEE, {kT, Infinity, n}]
```

$$\frac{3N}{\frac{2}{kT}} + \frac{aeF^{3/2}N\sqrt{\frac{1}{kT}}}{2\sqrt{2\pi}} + \left(-\frac{a^2eF^3N}{3\pi} + \frac{8a^2eF^3N}{27\sqrt{3\pi}}\right)\left(\frac{1}{kT}\right)^2$$

$$+ \left(-\frac{a^3eF^{9/2}N}{9\pi^{3/2}} + \frac{8\sqrt{\frac{2}{3}}a^3eF^{9/2}N}{27\pi^{3/2}} - \frac{a^3eF^{9/2}N}{9\sqrt{2}\pi^{3/2}}\right)\left(\frac{1}{kT}\right)^{7/2} + 0\left[\frac{1}{kT}\right]^{9/2}$$

The Maxwell (classical) case is found by setting $a = 0$

```
EnergyMaxwell = Normal[SEEE] /. a → 0
```

$$\frac{3kTN}{2}$$

The Fermi case is found by setting $a = 1$

```
EnergyFermi = SEEE /. a → 1
```

$$\frac{3N}{\frac{2}{kT}} + \frac{eF^{3/2}N\sqrt{\frac{1}{kT}}}{2\sqrt{2\pi}} + \left(-\frac{eF^3N}{3\pi} + \frac{8eF^3N}{27\sqrt{3\pi}}\right)\left(\frac{1}{kT}\right)^2$$

$$+ \left(-\frac{eF^{9/2}N}{9\pi^{3/2}} + \frac{8\sqrt{\frac{2}{3}}eF^{9/2}N}{27\pi^{3/2}} - \frac{eF^{9/2}N}{9\sqrt{2}\pi^{3/2}}\right)\left(\frac{1}{kT}\right)^{7/2} + 0\left[\frac{1}{kT}\right]^{9/2}$$

The Bose case is found by setting $a = -1$

```
EnergyBose = SEEE /. a → -1
```

$$\frac{3N}{\frac{2}{kT}} - \frac{(eF^{3/2}N)\sqrt{\frac{1}{kT}}}{2\sqrt{2\pi}} + \left(-\frac{eF^3N}{3\pi} + \frac{8eF^3N}{27\sqrt{3\pi}}\right)\left(\frac{1}{kT}\right)^2$$

$$+ \left(\frac{eF^{9/2}N}{9\pi^{3/2}} - \frac{8\sqrt{\frac{2}{3}}eF^{9/2}N}{27\pi^{3/2}} + \frac{eF^{9/2}N}{9\sqrt{2}\pi^{3/2}}\right)\left(\frac{1}{kT}\right)^{7/2} + 0\left[\frac{1}{kT}\right]^{9/2}$$

Appendix 4

Evaluation of the Correlation Function Integral

A.4.1. *Initial domain of integration*

In Sec. 5.2.1 we made use of the integral identity

$$\int_0^t d\tau_1 \int_0^t d\tau_2 G(\tau_1 - \tau_2) = 2 \int_0^t (t - \tau)G(\tau)d\tau,$$

which, we stated, was obtained by an appropriate change of variables. In this Appendix we shall show how this result is evaluated.

The integral

$$I = \int_0^t d\tau_1 \int_0^t d\tau_2 G(\tau_1 - \tau_2)$$

is a two-dimensional integral whose domain of integration is

$$0 < \tau_1 < t, \quad 0 < \tau_2 < t.$$

In the $\tau_1 - \tau_2$ plane this is the square with vertices A, B, C, D, shown in Fig. A.4.1.

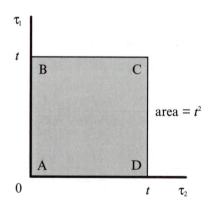

Fig. A.4.1.   Original domain of integration.

A.4.2. *Transformation of variables*

The argument of the integral is $\tau_1 - \tau_2$; this is constant along the lines $\tau_1 = \tau_2 + \text{const}$. It thus makes sense to change variables to

$$\tau = \tau_1 - \tau_2$$
$$T = \tau_1 + \tau_2$$

and the domain of integration in terms of the new variables may be found
from the location of the vertices:

	T_1	T_2	τ	T
A	0	0	0	0
B	t	0	t	t
C	t	t	0	$2t$
D	0	t	$-t$	0

Thus the domain of integration in the $T - \tau$ plane is as shown in Fig. A.4.2.

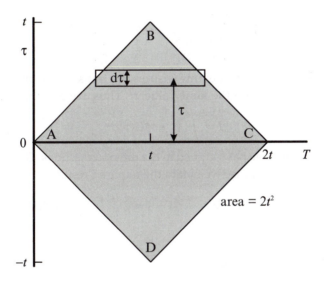

Fig. A.4.2.   Transformed domain of integration.

In terms of the transformed variables the argument of the integral is $G(\tau)$
and this is independent of the T variable. So $G(\tau)$ is constant throughout
each slice shown in the figure. Now the slice at height τ has width $2(t - \tau)$
and so its area is $2(t - \tau)\mathrm{d}\tau$.

A.4.3.   *Jacobian of the transformation*

The contribution to the integral is then $2(t - \tau)G(\tau)\mathrm{d}\tau$. However, we need
to multiply this by the Jacobian J of the transformation of variables. Thus

the integral is

$$I = 2 \int_{-t}^{t} (t - \tau) G(\tau) J \, d\tau.$$

The Jacobian is needed since the area of the transformed domain is double that of the original whereas the value of the integral must be independent of the transformation. Consider the case where $G(\tau) = 1$. Then

$$I = \int_0^t d\tau_1 \int_0^t d\tau_2 G(\tau_1 - \tau_2) = t^2,$$

the original domain area. The Jacobian is given by

$$J = \begin{vmatrix} \dfrac{\partial \tau_1}{\partial \tau} & \dfrac{\partial \tau_2}{\partial \tau} \\ \dfrac{\partial \tau_1}{\partial T} & \dfrac{\partial \tau_2}{\partial T} \end{vmatrix} = \begin{vmatrix} \dfrac{1}{2} & -\dfrac{1}{2} \\ \dfrac{1}{2} & \dfrac{1}{2} \end{vmatrix} = \frac{1}{2},$$

precisely the required value; this ensures the transformed integral in the case where $G(\tau) = 1$ has the same value, t^2. Thus, we conclude that

$$I = \int_{-t}^{t} (t - \tau) G(\tau) d\tau.$$

Usually, as indeed we have argued in the previous sections, $G(\tau)$ is an even function. And in that case we obtain the required result

$$\int_0^t d\tau_1 \int_0^t d\tau_2 G(\tau_1 - \tau_2) = I = 2 \int_0^t (t - \tau) G(\tau) d\tau.$$

References

[1] J. W. Gibbs, On the equilibrium of heterogeneous substances, *Trans. Conn. Acad.* **3** (1876), 108–248; *Trans. Conn. Acad.* **3** (1878), 343–524; Reprinted in *The Collected Works*, Vol. 1 (particularly pp. 87–88), pp. 55–349 (Longmans, NY, 1931), P. Duhem, *Le Potentiel Thermodynamique et ses Applications* (Hermann, Paris, 1886).

[2] H. B. Callen, *Thermodynamics and an Introduction to Thermostatistics*, 2nd ed. (John Wiley, 1985).

INDEX